THE END OF FOOD

THE END OF FOOD

Thomas Pawlick

BARRICADE
BOOKS
Fort Lee, New Jersey

Published by Barricade Books Inc.
185 Bridge Plaza North
Suite 308-A
Fort Lee, NJ 07024

www.barricadebooks.com

Copyright © 2006 by Thomas Pawlick
All Rights Reserved.

Library of Congress Cataloging-in-Publication Data
A copy of this title's Library of Congress Cataloging-in-Publication Data is avail-
able upon request from the Library of Congress
ISBN 1-56980-302-1
Printed in Canada

First Printing

Contents

THE END OF FOOD

Part I : THE PROBLEM

Red Tennis Balls

1

THE TOMATO WAS THE LAST STRAW. That's sort of a mixed metaphor, but how else can you say it?

I wanted to make a salad, a simple thing, just lettuce, tomatoes, cucumbers, some parsley, add a can of tuna, and toss it in vinegar and oil: a quick meal, so I could get to work on the stuff I'd brought home from the office.

But when I went to slice the tomato, it was too hard.

Red, but too hard for eating. A tomato should be just starting to get soft and juicy for the flavor to be there. Hard tomatoes, unless you're frying them green southern-style, are bland and tasteless. They shouldn't crunch when you eat them.

Okay, pick another one out of the batch. I'd bought four at the supermarket the day before and one of the others would be ripe.

I squeezed the second one. It was hard too. So were the third and the fourth. I looked at them. They were all bright red, not green. Yet they seemed nearly as tough and crunchy as so many raw potatoes.

Oh well. Put them back on the counter. In a day or two they'd be ripe enough.

But a day or two later, they weren't.

A week later, they were still hard.

So I put them on the windowsill, directly in the sun, to ripen. Two, three days went by, then a week.

Still hard.

This isn't possible, I thought. *Tomatoes are supposed to ripen in the sun. They are supposed to get soft and juicy so that when you slice them for your salad they taste yummy.*

Not these tomatoes.

I'm a stubborn man. I was determined that no mere vegetables (actually tomatoes are classed as "berry fruits") would get the best of me. I would outwait them, at least for a little longer.

But it did no good. They were as red as little fire engines when I'd picked them off the supermarket shelf and taken them home. But all these days later they were still not ripe. They *looked* ripe. No tomato could look riper. But that was all. They were not soft and juicy, rich with flavor as they ought to be after all that sunlight and patience.

One had a tiny dark spot, where some sort of rot may have been starting, and another had mold around the mark where the stem broke off, but it still wasn't soft.

Frustrated, I took one and went outside, where a wooden board fence separated the back patio of my city condo from my neighbor's. *You hard little devil*, I thought. *I'm going to do with you what vaudeville audiences used to do with tomatoes.* I threw it against the fence, expecting it to splatter like tomatoes used to splatter on singers with off-key voices and stand-up comics who told bad jokes.

It bounced off, undamaged, like a not-very-springy, red tennis ball.

I picked it up and threw it again, this time hard. It hit the fence, cracked—but did not splatter. When I looked inside, I found that some of the seeds inside the tomato had actually sprouted.

And that was the last straw, or tennis ball, or tomato.

My attention was well and truly fixed. As Sherlock Holmes would put it, the game—in this case the berry fruit—was afoot. Mixed metaphor number two.

A SHELTERED LIFE

By way of explanation here, so you won't think of me as a fanatic: At the time I met these rubbery North American supermarket tomatoes I'd just returned from six years of living in Italy, where the tomato is a kind of dietary deity. Although supermarkets exist in places like Rome or Milano, they—or at least their fresh produce sections— haven't really caught the average food-worshiping Italian's fancy. Most people still buy fruit and vegetables in open, neighborhood farmers' markets, where produce is displayed in mind-boggling variety under colorful awnings, out in the sunny piazza. The quality is . . . *mama mia!* Who can describe?

Before that our family lived on a small farm in rural Canada, where we raised our own home-garden tomatoes, picked them off the vine, and ate them fresh in the field.

It had been a sheltered life. This episode with the tennis balls was a shock.

On Saturday, I went back to the supermarket and gave the tomato shelf a closer look. There were three bins of tomatoes: one labeled "field grown," another "greenhouse," and a third full of elongated pasta tomatoes. The greenhouse tomatoes looked so exactly alike—all the same size, perfectly round and all exactly the same, uniform red color—that it was almost unreal, like maybe they were wax tomatoes cast from the same mold for decorative display.

But they weren't wax.

The pasta tomatoes also looked exactly like each other: same size, same shape, precisely the same color, like tiny Italian soldiers in red uniforms.

Only the field tomatoes had any differences. They were the same size and shape, but some had a slight hint of yellow or green around the stem scar, where others were totally red. I bought some of the greenhouse and some of the field tomatoes, and took them home.

Both types were hard, and by now I knew better than to hope for any changes. I cut them up for salad, the knife crunching through their tough outer walls (what botanists call the pericarp wall). I was surprised by how thick these were, and went for a ruler: more than three-eighths of an inch. In a later trip, to a different supermarket, I found one with walls just short of half an inch thick. The greenhouse tomatoes tended to be thinner.

When I ate them ... well, let's just say they tasted vaguely like tomatoes.

But comparing them to what the Romans eat, or to what our garden used to produce every summer, was like comparing carbonated cat pee to a rich, foaming Guinness stout. They were from a different planet. A hard, red, rather bleak planet.

Why? What had happened to these tomatoes?

It took some searching—through libraries, the Internet and the horticulture faculties of several universities—but eventually a story came together: the story of the slow ruination not only of the North American tomato, but of most of the good, tasty, nourishing food of all kinds that Americans and Canadians once ate, and took for granted.

But more on that later. Let's stick with tomatoes for now.

FOOD TABLES

Periodically, the United States Department of Agriculture (USDA) publishes a set of tables, generally referred to as "food tables," although different versions have had different formal titles. These list a wide variety of foods, from meat, fish, and grain to fruits and vegetables, giving measures of their actual content in terms of vitamins, minerals, protein, and other substances that can play a part in the human diet. Researchers and specialists in nutrition refer to them frequently as a base measure for making general comparisons.

The first of these was put out by the USDA's Dr. W. O. Atwater, a pioneering food researcher assisted by fellow scientist Charles D. Woods, under the title *The Chemical Composition of American Food Materials* (1896)[1]. This early work, completely superseded by later research, looked at a relatively small number of foods, and tested for an even smaller number of ingredients. Only six measures in it compare roughly with later, much more detailed tables for untreated (namely, not peeled, canned, or otherwise processed) tomatoes. As Atwater's sample sizes may have been different than those used today, accurate comparison is actually impossible.

Later USDA publications, however, generally give figures for either "100 grams of edible portion" of the food in question, or for the nutrients in one pound of a given food item. Thus, they can be compared. And comparisons–even between fairly recent tables–are more than enlightening. They are shocking.

The most recent set of tables, posted on the USDA website for 2002, is titled *USDA National Nutrient Database for Standard Reference, Release 15.*[2] Comparing the figures in it with those given in *USDA Agriculture Handbook No. 8: Composition of Foods*, published in 1963,[3] shows that 100 grams of today's average red, ripe whole tomato contain 22.7 percent *less* protein than a tomato would have if purchased by American shoppers in the year President John Kennedy was murdered in Dallas.

The main reason for including tomatoes in our diets, of course, is not for their raw protein value, but because they are normally supposed to be rich in vitamins A and C, in potassium, phosphorus, iron, and calcium, as well as in the possibly cancer-suppressing lycopene. Lycopene is a "carotenoid," the group of yellow, orange, and red plant pigments that give carrots, watermelons, and tomatoes their colors. Some carotenoids are "precursors" of vitamin A, which is to say they help produce it as a result of chemical reactions in human organs like the liver. While lycopene isn't an actual vitamin A precursor, it is a powerful antioxidant that "seems to inhibit the reproduction of cancer cells."[4] Unfortunately, the USDA food tables don't measure lycopene.

They do measure vitamin A, however, a nutrient which is needed by the human body to maintain good eyesight as well as normal sexual reproductive health and body growth. And they measure vitamin

C, required to prevent a variety of diseases, from scurvy to the common cold, to control stress, to maintain normal arteries, and to help heal cuts and wounds.

Tomatoes were once among the best sources of these vitamins. But 100 grams of today's fresh tomato contain *30.7 percent less Vitamin A and 16.9 percent less Vitamin C (ascorbic acid) than its 1963 counterpart.* It also has fully 61.5 percent less calcium (required to maintain strong bones and teeth), 11.1 percent less phosphorus, 9 percent less potassium, 7.97 percent less niacin, 10 percent less iron and 1 percent less thiamin.

And those are only the losses since 1963, about half a person's lifetime. If today's values are compared with those for earlier years, the story is often worse. The amount of iron in 100 grams of raw red tomato today is 10 percent less than in 1963, but *fully 25 percent less than in 1950,* when the real-life counterparts of the characters in television's M*A*S*H were busy fighting the Korean War.[5] The amount of vitamin A, measured in International Units (IUs), is 43.3 percent less than in 1950. How much has been lost since 1930, or earlier? No one can say. But a trend in losses of key nutrients is obvious.

Of course, not every substance found in fresh tomatoes has diminished. Two, in particular, have posted spectacular increases since 1963. The amount of fat (lipids) has climbed by 65 percent, while sodium—the basis of common table salt (sodium chloride, or NaCl) —has leaped upward by an astounding 200 percent.

These increases and decreases are not isolated, but can have a kind of domino effect, mutually reinforcing each other. For example, sodium (as sodium chloride), has for years been considered the primary factor responsible for high blood pressure. As the authors of *Understanding Nutrition,* a basic college textbook for the health sciences, put it:

> *Some individuals respond sensitively to excesses in salt intake and experience high blood pressure. People most likely to have a salt sensitivity include those with chronic renal disease, diabetes, or hyper-*

*tension, African-Americans, and people over 50 years of age.
Overweight people also appear to be particularly sensitive to the
effect of salt on blood pressure. For them, a high salt intake correlates
strongly with heart disease and death.*[6]

The authors also note that a high sodium intake can be linked to
the amount of calcium in the human body—a factor that may be crucial in the development of osteoporosis, the so-called "brittle bones"
disease of the elderly. Sodium appears to have a negative influence on
how much calcium is retained by the human body. "Dietary advice to
prevent osteoporosis might suggest eating more calcium-rich foods
while eating fewer high-sodium foods," warn the nutrition textbook's
authors.[7]

And what have tomatoes lost since 1963? Fully 61.5 percent of
their calcium. What have they gained? Two hundred percent sodium.

The picture becomes still more interesting when one looks at the
7.97 percent loss in potassium. "Low potassium may be as significant
as high sodium when it comes to blood pressure regulation," says the
nutrition textbook.[8] And it adds: "Even when potassium isn't lost, the
addition of sodium still lowers the potassium-to-sodium ratio.
Limiting sodium intake may help in two ways then—by lowering
blood pressure in salt-sensitive individuals and by indirectly raising
potassium intakes in all individuals."[9]

The modern fresh market tomato appears to be aimed at doing
exactly the opposite.

Higher in fat, higher in sodium, lower in calcium, potassium,
Vitamin A, and Vitamin C, losing iron, phosphorus, niacin and thiamin, today's tomato looks as if it is almost calculated to lack whatever nutritionists recommend.

Processed tomato products have suffered a similar fate. Since
1963, for example, canned tomato juice has lost 35.5 percent of its iron
and 30.5 percent of its vitamin A. Since 1950, the amount of vitamin
A in tomato juice has dropped 47 percent–almost by half. As for toma-

to catsup, it has lost 13.6 percent of its calcium, 12.5 percent of its iron and 27.4 percent of its vitamin A since 1963.

At the same time, sodium has increased 13.8 percent, and fiber (perhaps reflecting plant breeders' desire for those tough outer walls) has jumped upward by an amazing 1,200 percent.

Not only is the tomato losing beneficial nutrients, but its supermarket version is also losing in another key category: variety.

DIMINISHING CHOICES

It's not certain precisely how many varieties there are of *Solanum esculentum*, the Latin name of the common tomato (it used to be called *Lycopersicon esculentum*, which means "wolf peach," but has been renamed). Native to Latin America, and cultivated for centuries by the Indians of Mexico and Peru, it was adopted as early as 1554 by the Italians and by Americans (who were slow cluing into the delights of pasta sauce) in the early 1800s. Over the years, plant breeders have developed literally thousands of varieties, ranging from plants with big, fat yellow and orange fruit to tiny little red cherry tomatoes. Some have thin walls, some thick, some are sweeter, some less sweet, some ripen early, some late, some are more frost- or disease-tolerant, some less.

But the key word is choice. According to the Decorah, Iowa-based Seed Savers Exchange,[10] which caters to home gardeners, there are more than 5,500 varieties of tomato in its collection alone.[11]

How many of all those thousands of possible varieties show up in our supermarkets, either as so-called "fresh market" tomatoes or processed into tomato products ranging from pasta sauce to tomato paste, salsa, or catsup?

Not many. The North American supermarket system gets most of its tomatoes from only four locations. According to extensionist Dr. Tim Hartz, of the University of California at Davis, more than 85 percent of the tomatoes shipped for processing into canned or other products come from California.[12] The California Tomato Growers

Association likes to boast that "nine out of every 10 tomatoes processed in the U.S." come from that one state alone.[13]

As for fresh-market tomatoes, sold unprocessed as harvested from the field, the Florida Tomato Committee reports that more than 50 percent come from Florida.[14] During the December through May winter season, when states like Ohio or Virginia can't grow anything, Florida and California are the *only* states shipping tomatoes. Recently, due to the North American Free Trade Agreement (NAFTA), Florida has had to compete with Mexico during the winter months, but it still has the lion's share of the fresh market. In Canada, Florida also dominates the winter fresh market, although recently it has had some competition from Mexico and from such European Union countries as Spain and Portugal, both of which have major greenhouse tomato growing industries. In summer, Canada supplies some of its own tomatoes, mostly field- or greenhouse-grown in southern Ontario, especially near Leamington.

How many choices are available to the consumer, in terms of variety?

According to the Florida Agricultural Statistics Service, during the 1999-2000 growing season, 11 varieties dominated the fresh market, with only five accounting for more than 80 percent of all Florida tomatoes grown. The favorite, by far, was Florida 47, which accounted for 35.9 percent of all varieties grown.[15] In the EU, which exports to Canada in winter, the story is much the same. In Portugal in 1999, for example, more than 80 percent of the tomato crop was accounted for by only six varieties.[16]

In California in the year 2002, according to the Department of Vegetable Crops of the University of California at Davis, only 10 varieties accounted for more than 60 percent of the entire processing tomato market.[17] Five of these (nearly 26 percent) were proprietary varieties, developed by major multinational food processing companies that require their contract suppliers to grow only their in-house varieties.

If we take 6,000 or more as a very rough benchmark figure for the total number of North American tomato varieties known–and this is an almost ridiculously conservative number—the math is revealing. The 15 American-grown varieties that dominate both process and fresh market tomatoes available in our supermarkets today represent only 0.25 percent of the possibilities that could be out there. One quarter of one percent.

Some choice.

So, we have a minuscule number of tomato varieties available to shoppers, and a diminishing amount of nutrients in fresh tomatoes (with the exception of the rising amount of fat and sodium). These tomatoes look wonderful–big, bright red, perfectly round, unblemished, as uniform as if they'd been turned out with cookie-cutters. Yet they are tough and rubbery, and at least to my own purely subjective taste, comparatively flavorless.

How did this come to pass?

The answer is, by deliberate selection. The huge, multinational corporations that dominate the continental food industry, from seed to supermarket shelf, prefer it this way.

After days of searching and phone calls, I located several industry spokesmen and scientific experts on tomato breeding, including specialists in fresh market varieties and others focused on the process market. I spoke to some of them for a half hour or more, while the tape cassette turned, asking for detailed descriptions of the characteristics that made the top 15 tomato varieties such a success in their respective markets.

As one scientist said, "the first characteristic is yield, the second is yield, and the third is yield." He was, of course, being facetious.

According to the scientists, the characteristics that make a tomato variety a hit in the fresh market category include, in order of importance:

1. yield (in pounds per acre)
2. large size (200-250 grams)

3. firmness, in terms of thickness and hardness of the outer pericarp wall (which provides the ability to withstand pressure and between 25,000 and 50,000 pounds of weight when bouncing along in a truck during shipment)
4. resistance to disease
5. heat tolerance (in setting fruit during Florida's warm weather)
6. uniformity of shape
7. uniformity in time of ripening (color)

I asked one expert if any other characteristics were desirable. He paused for a moment to think, then concluded: "no, you've got quite a bit of it."

Processing tomato experts had a somewhat different list. In California, the top characteristics were:

1. yield (in pounds per acre)
2. viscosity or thickness (which governs how much of a product can be made from a pound of the given tomato's paste)
3. amount of soluble and insoluble solids in the fruit
4. firmness (ability to withstand rough handling during mechanical harvesting)
5. uniformity of color
6. disease resistance
7. heat and cold tolerance (so as to continue producing at the early and late ends of the season)

I also asked these experts if any other characteristics were important, if we'd left any out. "No," said one. "It does get to be end product-driven."

No one mentioned the two characteristics that any ordinary consumer would likely put at the top of his or her list, namely: flavor and nutritional content. These were simply *not there*, not important, not even worth mentioning.

For the modern, corporate food industry—whose needs these university horticulturists' research serves—how a food item tastes and whether or not it is nourishing for human beings appear *not to be issues*. They aren't even discussed.

A quick survey of various websites where university horticultural researchers report the results of field trials of plant varieties gave the same result. There were reports of trials at a number of universities, mostly in the south and midwest. The typical list of qualities tested for in fresh market tomatoes included "yield, earliness, fruit size, fruit resistance to cracking, firmness, acidity, and plant tolerance/resistance to diseases."[18] Flavor and nutritional value were almost never mentioned.

From every indication, if a tomato variety were developed that was perfectly uniform in shape and size, that grew fast in both heat and cold, that ripened at exactly the same moment every season, that had an outer layer as tough as brake lining, and that yielded massive amounts per acre–but which had no flavor whatsoever, and absolutely no nutritional value–the industry would likely welcome it like the Second Coming of Christ. And consumers, you and I, would be expected to buy the things or just bloody well go hungry.

Of course, even if we bought them and ate them, we would still be "going hungry," because we'd have received very little nutrition.

But hey, that's capitalism, eh?

As for flavor, I wasn't just imagining that the fresh market supermarket tomatoes I'd bought were less tasty than those I'd eaten in Italy, or taken from my home garden years ago. According to the textbook *Economic Botany: Plants in Our World*:

> *Tomatoes to be sold as fresh grocery store produce are picked before they are mature or when they are only beginning to turn colorful and then are ripened at the time of selling. Fruits that have been picked green are tough because of a lack of the proper ethylene-generated maturation, or they are mushy because the intercellular matrix dete-*

riorates. They also lack the sugar that accumulates very rapidly at the peak of ripening when tomatoes are left on the vine....Ethylene [is] a plant hormone that is responsible for the series of events that lead to the final color change, softening, and flavor production characteristic of natural ripening. [19]

So, not only were my store tomatoes deliberately bred to be tough to withstand the bouncing of long range transport, they were rendered still tougher by picking them when green. Tough and tasteless.

By "ripened at the time of selling," the textbook was referring to the practice of artificially gassing the green tomatoes with ethylene during or just after transport in special "ripening rooms." This gives them a suddenly red color, making them look good on the shelf, but doesn't appear to have the same effect as natural ripening in terms of producing flavor and texture.

Refrigeration during transport has an even more negative effect. According to a USDA Agricultural Research Service study of the effects of refrigeration on tomato flavor, "chilling the fruit reduced ripe aroma, sweetness, and general tomato flavor, while increasing sourness and reducing sweetness. This was supported by measured changes in aroma compounds, sugars, and acids."[20]

Those little red tennis balls had begun to educate me. They had also begun to really annoy me. I wanted to know more about the system that was doing this. I wasn't about to take it lying down.

Neither should you. And you don't have to. There are alternatives, which we will discuss in later chapters, after some of the other aspects of our modern, corporate North American food system have been described.

The End of Food | 2

THE DECLINE OF THE TOMATO ISN'T THE ONLY tragic story in the modern supermarket, nor are the USDA's nutrient tables the only source documenting what's happening. A tour through the recent literature on food in the U.S., Canada, and Britain, followed by a walk through just about any corporate chain-owned grocery store, produces a horrific picture of losses so steep, and continuing at such a rate, that it is not an exaggeration to speak–literally–of the coming end of food.

Start with some of the most obvious losses, and their potential consequences in terms of human health.

On July 6, 2002, the Toronto *Globe and Mail* began publishing a series of articles on food, including one by reporter Andre Picard, who wrote:

> *Fruits and vegetables sold in Canadian supermarkets today contain far fewer nutrients than they did 50 years ago. Vital vitamins and minerals have dramatically declined in some of our most popular foods.*
>
> *Take the potato, by far the most consumed food in Canada. The average spud has lost 100 percent of its vitamin A, which is impor-*

tant for good eyesight; 57 percent of its vitamin C and iron, a key component of healthy blood; and 28 percent of its calcium, essential for building healthy bones and teeth.

It also lost 50 percent of its riboflavin and 18 percent of its thiamine. Of the seven key nutrients measured, only niacin levels have increased.... The story is similar for 25 fruits and vegetables that were analyzed [in a Globe and Mail, CTV *study].*[1]

Picard's numbers were based on food tables supplied by the Canadian government, but were fairly close to what the USDA tables were showing. In fact, some of the Canadian data was taken originally from the USDA tables. What Picard was saying was generally true for the U.S. as well.

It was also true in Britain. There, researcher Anne-Marie Mayer published a study in the *British Food Journal*, a respected scholarly source for nutritionists and food specialists. She wanted to answer the question: "Has the nutritional quality (particularly essential mineral content) of fruits and vegetables changed this century during the period of changes in the food system and the modernization of agriculture?"[2] To get her answer, she looked at the United Kingdom's equivalent of the USDA food tables, *The Chemical Composition of Foods*, for 1936 and 1991, and compared the contents of 20 vegetables and fruits. Her results:

There were significant reductions in the levels of calcium, magnesium, copper and sodium in vegetables, and magnesium, iron, copper, and potassium in fruits. The greatest change was the reduction in copper levels in vegetables to less than one-fifth of the old level. The only mineral that showed no significant differences over the 50-year period was phosphorus.[3]

SMALL NUMBERS, BIG CONSEQUENCES

Numbers. Most supermarket shoppers are not math majors, nor

experts in statistics. And besides, numbers sitting there on the page all by themselves are a little boring. Just how important is it whether the trace amounts of copper, say, in a given food are going up or down by a few milligrams? A milligram is a very small thing, one one-thousandth of a gram. And a gram is only a bit more than a twenty-eighth of an ounce. A milligram of some substances would be so small we couldn't even see it without a microscope–a mere speck. What difference does it make whether we swallow a couple of specks of anything on a given day?

The answer is, a big difference. The human body is a mysterious thing, resilient and adaptable, but amazingly complex. The very resilience and adaptability that have made us the most successful species in the history of our planet depend on an intricate number of finely tuned relationships–between our bodies and the environment, between each of the organs inside our bodies, and between each of the cells inside those organs. We walk a thousand minute bioenvironmental tightropes every day, every moment, teetering and tipping, and righting ourselves just in time to maintain equilibrium, staying within the narrow band of conditions, both internal and external, that allow us to survive.

If the outside temperature falls below 32 degrees F and we aren't wearing protective clothing, we freeze to death in a few minutes. If it climbs much above 120 degrees F, and we can't find shade or an air-conditioned building, we can die of heatstroke.

I've personally experienced both temperature extremes. Stranded by the highway late at night, during a January blizzard in Quebec, I huddled inside my snowbound car with only a thermal snowmobile suit and a lit candle between me and the –28 degree F winter wind outside. When the snowplow smacked into my back bumper in the morning, I was grateful that a couple of spots of frostbite on my cheeks were the worst the cold had done.

A decade later, crossing the Arabian desert during the first Gulf War, with temperatures well above 120 degrees F, I noticed that the air actually burned my lungs when I inhaled, while the sweat evaporated almost before it formed on the surface of my skin. It was a tremendous physical relief to pull off the road and walk into an air-conditioned restaurant in that throbbing, shimmering heat. If I'd stayed out in it much longer, after five hours in a car whose air-conditioning system had broken down, I'd have paid a steep price.

A narrow band.

As for copper, we're all familiar with this practical, beautiful metal. We make pennies out of it, and electrical wiring, and polished kettles to boil our tea water. Anyone who's traveled to Michigan's Upper Peninsula has seen copper ore, and probably bought a nugget or two at some souvenir stand. Copper roofs, turned green by contact with the air, adorn many of our best-known architectural monuments.

A normal, healthy person has about 100 milligrams of copper in his or her body, distributed throughout a variety of cells and tissues. Only 100 mg. A few specks. What can happen if that number goes up or down?

An article in a recent issue of *Discover* magazine ought to give an idea.[4] The author, a neurologist from Concord, New Hampshire, recounted his experience with a 22-year-old patient he called Megan, who suffered from a disorder called Wilson's disease. Wilson's is a genetic disorder that prevents the body from properly eliminating excess amounts of copper. As Dr. John R. Pettinato explained:

> Copper is an essential trace element, and most diets provide about one quarter more than is needed for cellular metabolism. The liver processes this excess copper into bile, which is excreted in the stool. Some people inherit a defect in this processing pathway, and symptoms occur as harmful amounts of copper accumulate in the brain and the liver.[5]

A few specks too many accumulated in Megan. She became depressed, anxious, and developed anorexia, as well as a bad case of the shakes. Her legs and head shook, and she was rarely tremor-free. Then she began to drool, especially at night; "her extremities had become stiff, and her arms didn't swing naturally when she walked. She felt dizzy and off balance and seemed to shuffle."[6]

If Pettinato hadn't quickly diagnosed and treated her, Megan might have gotten a lot worse. The full range of symptoms of Wilson's can include hepatitis, liver damage, tremors, slurred speech, lack of coordination, cramping, emotionality, depression, parkinsonism, psychosis, and "other bizarre behaviors." Some patients die. All that from a few specks too many of a single element.

Of course, Megan was suffering from too much copper. What about too little? As the authors of *Understanding Nutrition* note, copper deficiency is relatively rare, but is seen in some malnourished children. "Copper deficiency in animals raises blood cholesterol and damages blood vessels, raising questions about whether low dietary copper might contribute to cardiovascular disease in humans," say Whitney and Rolfes.[7]

Copper is only one of the many nutrients our bodies need. Some, like iron, or vitamins A and C, are of major importance and have been studied in great detail over the years. Others, like selenium or molybdenum, or vitamins E or K, have received less attention and their roles in keeping us healthy are only beginning to be understood. As late as 1975, the USDA food tables didn't even list selenium or vitamins D and E, and they have only recently begun to include the amino acids.

Also only partially understood are the effects each of these nutrients have on each other, or on the body when working in tandem, such as the interconnections between sodium and calcium intake noted in Chapter One. The point is that they are *all* important; each one affects the others, working with them or against them, in an intricate living symphony of chemical and biochemical reactions. Even the smallest excesses or deficiencies can provoke myriad unexpected results, which we ignore at our peril.

SCURVY KNAVES

In the 1800s, when Herman Melville wrote his classic whaling novel *Moby Dick* (the movie version, a century later, starred Gregory Peck), sailors would stay at sea for months, even years, and their stores of fresh vegetables would often be exhausted long before they could put into port for more provisions. Forced to subsist on diets of salt pork and biscuit, they developed a whole range of diseases stemming from dietary deficiencies, the best known of which was scurvy ("Ahoy there, you scurvy knave!").

The first sign of scurvy was fatigue, which kept getting worse. Then the sailor's gums would start bleeding, followed by his skin. The blood vessels under his skin would appear to turn red and swell. If the man cut himself, the cut wouldn't heal. His fingers and toes would swell, and his body hair would turn curly and kinky. Horny growths would appear on the skin, particularly his buttocks. He would experience increasing pain in his joints, would become pale and lethargic and unable to sleep. Next his teeth would start falling out and finally he would start to hemorrhage. Finally, thankfully, he would die.[8]

As many as two-thirds of a ship's crew would die this way during a long voyage. Experiments by British physician James Lind finally isolated the cause—lack of citrus or other fruits containing what was then called the "antiscorbutic factor." Isolated nearly 200 years later, the factor was found to be a carbon compound similar to glucose, which was dubbed "ascorbic acid"—today's vitamin C.[9] Eventually, the British navy solved the problem by requiring all of its sailors to drink lime juice during long voyages, thus giving rise to the slang nickname for an Englishman, "limey."

And what has the potato lost over the past 50 years? In Canada, 57 percent of its vitamin C. The American tomato has lost 16.9 percent of its vitamin C just since 1963. And broccoli, described by reporter Picard as "a food that epitomizes the dictates of healthy eating,"[10] has, according to the USDA tables, lost fully 45 percent of this crucial nutrient since John Kennedy died.

Are Americans and Canadians likely to break out suddenly with the symptoms of advanced scurvy? Probably not in the short-term future, since other food items—including limes, lemons, and grapefruit—still contain considerable ascorbic acid. But the general trend toward drastic vitamin C loss in so many food items at the same time, a steady move away from what makes for good health and toward nutritional poverty, is hardly reassuring.

It's even less reassuring if one takes such scourges as heart disease or cancer into account. The causes of cancer, what John Wayne called "the big C," and which killed him shortly after he made his last classic western, *The Shootist*, are in many ways still a mystery to researchers. But so-called "free radicals" are not that mysterious.

As most of us were probably told in high school chemistry class (and promptly forgot once the exam was over), free radicals are molecules, or groups of atoms, in which one of the atoms in the group has an "unpaired" electron in its outer shell, making it unstable. Since atoms always seek stability, these molecules only exist very briefly, as intermediate products of earlier chemical reactions. As soon as they encounter another molecule with which they can combine, or from which they can scavenge an electron to pair with their extra one, they do so.

The human body is constantly creating free radicals, most often during the process of oxidizing, or "burning" food for energy.[11] This process produces a type of free radical called "reactive oxygen," which can begin a very destructive chain reaction as it attempts to bond with other atoms and achieve stability. It's a bit like the biblical raging lion, which "goes about, seeking whom it may devour." Snatching an electron from another atom, leaving it unstable, the oxygen radical creates "another, uglier than itself," which will in turn attack yet other atoms, creating more free radicals, and so on, and so on.

Raging about within our bodies, these sub-microscopic biochemical lions "may irritate or scar artery walls, which invites artery-clogging fatty deposits around the damage," the so-called hardening of the arteries that leads to heart disease.[12] There is also "a growing body of evidence" that "many of the things we associate with getting older—memory loss, hearing impairment—can be traced to the cumulative effects of free radicals damaging DNA ... thus diminishing the body's energy supply."[13] Scientists have also implicated oxidative stress in the development of arthritis and cataracts.

Worst of all, free radicals can have a "mutation-causing or mutagenic effect on human DNA, which can be a factor leading to cancer."[14] Too many free radicals, in fact, may have been what killed "the Duke," the very symbol of cowboy courage and manly strength.

A normal, healthy human body provided with a balanced diet has a set of natural defenses against free radicals, in the form of "anti-oxidants." These are substances which can chemically interact with free radicals and "neutralize" them in various ways without themselves turning into radicals. Like so many microscopic Buffys stamping out vampires without becoming vampires, they de-fang the radicals, rendering them harmless.

Foremost among these are various enzymes (proteins that help along chemical reactions without themselves being changed in the process), and the vitamins C and E. The chemical "de-fanging" activity of the enzymes depends heavily on the presence of the minerals selenium, copper, manganese, and zinc.

And exactly what is missing or declining in the foods sold in our modern supermarkets? Vitamin C (decreased by 57 percent in Canada's potatoes, declining fast in America's tomatoes, broccoli, and a host of other vegetables and fruits), and copper (down across-the-board by four-fifths in vegetables in England, but unfortunately not measured in the USDA tables for 1963 or 1975). The USDA did not analyze for selenium, manganese, or zinc until recently, nor for vitamin E.

What about vitamin A, down by 43.3 percent in red, ripe tomatoes in the U.S. since 1950, by 30.5 percent in tomato juice and 27.4 percent in tomato catsup since 1963? What is it good for?

First, it plays a crucial role in vision, helping to maintain the clarity of the cornea of the human eye, and in the conversion of light energy into nerve impulses in the retina. Without sufficient vitamin A, people can go blind.

In addition, vitamin A is needed to maintain what biologists call "epithelial" tissues in the body. These are the cells that form the internal and external surfaces of our bodies and their organs. They include our skin, which shields us from the outside world, and the walls that separate each of our internal organs from the others, as well as the mucus secretions that ease the movement of foods through the human digestive tract.

What could happen if a person were to stop eating vitamin A-rich foods? The authors of *Understanding Nutrition* are blunt: "Deficiency symptoms would not begin to appear until after [the body's] stores were depleted—one to two years for a healthy adult but much sooner for a growing child. Then the consequences would be profound and severe."[15]

In children, this could mean an upsurge in the negative effects of such infectious diseases as measles which, despite a vaccine available mostly in the rich countries, still kills some two million children worldwide every year. As Whitney and Rolfes explain: "The severity of the illness often correlates with the degree of vitamin A deficiency; deaths are usually due to related infections such as pneumonia and severe diarrhea. Providing large doses of vitamin A reduces the risk of dying from these infections."[16]

More obvious results would include night blindness, in which a vitamin A-deficient person's ability to see at night is sharply curtailed. Whitney and Rolfes provide a graphic description:

The person loses the ability to recover promptly from the temporary blinding that follows a flash of bright light at night or to see after the lights go out. In many parts of the world, after the sun goes down, vitamin A-deficient people become night-blind: children cannot find

their shoes or toys, and women cannot fetch water or wash dishes.
They often cling to others, or sit still, afraid that they may trip and
fall or lose their way if they try to walk alone.[17]

This condition may progress to total blindness.

Another result of vitamin A deficiency is "keratinization," a condition where the victim's epithelial surfaces are adversely affected. Mucus secretion drops, interfering with normal absorption of food along the digestive tract, causing general malnutrition. Problems also develop in the lungs, interfering with oxygen absorption, as well as in the urinary tract, the inner ear, and for women in the vagina. On the body's outer surface, "the epithelial cells change shape and begin to secrete the protein keratin—the hard, inflexible protein of hair and nails. The skin becomes dry, rough, and scaly as lumps of keratin accumulate."[18]

An attractive picture, eh? Blind, disease-prone children, short of breath, suffering from malnutrition, with problems peeing, and with scaly lumps all over their skins. Will we be seeing this in the near future? Again, probably not. But the tendency is there, and steadily increasing. Who can say where we'll be in another 20 or another 50 years, if present trends continue unabated? The Canadian potato, remember, has already lost *all* of its vitamin A.

And what of iron, down by more than half in Canada's potatoes, by 10 percent in the American tomato, and by various amounts in many other fruits and vegetables?

Statistically, low iron is the world's most common nutrient deficiency, and is particularly dangerous for menstruating or pregnant women and for growing children. Iron is absolutely necessary for the proper maintenance of hemoglobin in the blood and myoglobin in the muscles. It helps both of these proteins carry and release oxygen, permitting the biochemical reactions that give us energy. As the authors of *Understanding Nutrition* explain, a series of events can be triggered by insufficient iron in the body, events which can ultimately lead to life-threatening anemia:

Long before the red blood cells are affected and anemia is diagnosed, a developing iron deficiency affects behavior. Even at slightly lowered iron levels, the complete oxidation of pyruvate is impaired, reducing physical work capacity and productivity. With reduced energy available to work, plan, think, play, sing, or learn, people simply do these things less. They have no obvious deficiency symptoms; they just appear unmotivated, apathetic and less physically fit.... A restless child who fails to pay attention in class might be thought contrary. An apathetic homemaker who has let housework pile up may be thought lazy.[19]

If the iron deficiency continues and worsens, it eventually leads to full-blown iron-deficiency anemia:

In iron-deficiency anemia, red blood cells are pale and small. They can't carry enough oxygen from the lungs to the tissues, so energy metabolism in the cells falters. The result is fatigue, weakness, headaches, apathy, pallor, and poor resistance to cold temperatures... The skin of a fair person who is anemic may become noticeably pale. [20]

Such a condition can be particularly damaging if it occurs in growing children.

At the same time that they are losing nutrients, other vegetables and fruits are also suffering a drastic decline in the number of varieties available to consumers. Just as the number of tomato varieties is sharply limited in the supermarket, so are those of potatoes and apples. As investigative journalist Brewster Kneen noted in his landmark book, *From Land to Mouth: Understanding the Food System*:

Even though there are 2,000 species of potato in the genus solanum, all the potatoes grown in the United States, and most of those grown commercially everywhere else, belong to one species, solanum tuberosum. Twelve varieties of this one species constitute 85 percent of the U.S. potato harvest, but the one variety favored by most processors, the Russet Burbank, is by far the dominant variety. By 1982, 40 percent of the potatoes planted in the United States were Russet Burbanks.[21]

To witness the poverty of choice among apples, just walk into the corner supermarket and look on the shelves. Most chain stores have only three varieties on display: red delicious, golden delicious, and Granny Smith. Sometimes a Canadian store will also feature MacIntosh. Looking at such a display, it is useful to keep in mind that at the turn of the last century "there were more than 7,000 apple varieties grown in the United States. By the dawn of the twenty-first century, over 85 percent of these varieties, more than 6,000, had become extinct."[22]

By the year 2000, 73 percent of all the lettuce grown in the U.S. was one variety: iceberg.[23]

ACROSS-THE-BOARD DEGENERATION

Examples of the rapid decline in nutrients in our foods are not limited to vegetables and fruits. A general, across-the-board degeneration affects nearly everything we eat.

For instance, according to the USDA tables, chicken–which many of us eat in an attempt to avoid steroid-rich red meats–is in deep trouble. Skinless, roasted white chicken meat has lost 51.6 percent of its vitamin A since 1963. Dark meat has lost 52 percent. White meat has also lost 39.9 percent of its potassium, while dark meat has lost 25.2 percent.

And what has chicken gained? Light meat, 32.6 percent fat, and 20.3 percent sodium; dark meat, 54.4 percent fat and 8.1 percent sodium. Let's hear it for fat and salt.

Dairy products are no better. According to the USDA, creamed cottage cheese–eaten by millions of dieting men and women *precisely because* it is seen as a low-fat source of calcium and phosphorus to maintain strong bones and teeth–has in fact gained 7.3 percent fat since

1963, while losing 36.1 percent of its calcium, 13.1 percent of its phosphorus, and—incidentally—fully 53.3 percent of its iron. And what has it gained, besides fat? Hey, you guessed it: 76.85 percent in sodium.

We are also seeing increases in carbohydrates, which include sugars and starches. Good old healthy broccoli, for example, while losing 45 percent of its vitamin C, has seen its carbohydrate content jump upward by 13.8 percent since 1963.

As for bread, traditional mainstay of the Western diet, the highly processed nature of the typical soft, all-but-crustless, bleached-flour white supermarket loaf makes it hard to evaluate. In the process of manufacture, the nutritionally best parts of the original wheat grain are sifted, milled, or chemically bleached out of the flour to make it as perfectly white as possible. The purpose here is purely cosmetic, designed to accommodate the widespread–and completely irrational–public prejudice that white bread is somehow "better." Then a small portion of these nutrients is put back in to "enrich" (the pure irony of the industry's euphemism here is almost comic) what would otherwise be a loaf of nothing. So-called "enriched" white bread thus actually does contain some nutrients. But compared to loaves made by more traditional baking methods, the supermarket product is rather pathetic.

This was demonstrated as long ago as the 1970s, when the consumer-oriented *Harrowsmith* magazine conducted a comparison analysis of three loaves of bread: a) a mass-market "enriched" white loaf of Weston's bread, taken from the supermarket shelf; b) a white loaf made by a local, small-town bakery; c) a home-baked loaf that used the "Cornell bread" recipe developed by nutritionist Dr. Clive MacKay. The results of an independent laboratory analysis:

The Weston loaf proved highest in fat and chloride (calculated as salt) and lowest in protein and phosphorus, as well as in the B vitamins (niacin, thiamine, riboflavin and B_{12}). The homemade loaf was highest in protein, iron, calcium, phosphorus and the B vita-

mins niacin, riboflavin and B_{12}. It was lowest in chloride and, surprisingly, in fiber. The bake shop loaf was highest in fiber, in vitamin B_6 and folic acid, and lowest in both fat and calcium."[24]

Even those who represent the manufacturers of the spongy white "tissue bread" sold in supermarkets admit its inferiority. The managing director of the Baking Council of Canada, the lobbying and public relations arm of the baking industry, told *Harrowsmith*'s reporters he "does not eat Weston or any other mass-produced bread himself... he shops instead at a small specialty bakery ... adding that large industrial bakers could not match its quality."[25]

Then there is that traditional favorite, the hot dog, so closely identified with warm summer days and baseball, with Fourth of July and Canada Day picnics, as to be a virtual North American icon. No diehard fan's day at the diamond could be complete, sitting out in the bleachers, without a cold beer and a couple of hot dogs for lunch.

Except that the dog, in terms of food value, is almost worthless—especially the prized all-beef frankfurter, for which people are ironically willing to pay a premium, thinking they're getting something more for their money.

The average hot dog is actually 58 percent water, 20 percent fat, 3 percent ash and 6 percent sugar. Less than 13 percent of each sausage is made up of actual protein, and even this is of poor quality, consisting for the most part of scrapings from animals' bones after the main cuts of meat have already been taken in the packing plants.[26]

An eight-year study by the non profit *Protegez-vous* (protect yourself) magazine in Canada's Quebec Province found that most hot dogs–including all-beef, as well as blended chicken and turkey, and blended beef and pork hot dogs–contained "the minimum required by law of protein, too much sodium and too much fat" and that they were of generally "bad quality."[27] Only vegetarian hot dogs contained a reasonable amount of food value.

As for "all-beef" hot dogs, which many buyers seek on the mistak-

The End of Food : : :   29

en assumption that they contain more protein and less "fatty" meats like pork, the magazine concluded: "The all-beef hot dogs were the worst of our study: Not only are they among the most costly, but their saturated fat and sodium content is much too high. Avoid them."[28] The study authors advised readers that veggie dogs have "almost double the protein, a third less fat and sodium, and hardly any saturated fats compared to meat hot dogs."[29]

PILL POPPING TO THE RESCUE?

The above examples, taken chiefly from the produce, meat, and dairy sections, don't even touch on the subject of the highly processed foods that make up most of the other products found in the cans, heat-sealed tinfoil or plastic envelopes, and cardboard boxes that line an increasing percentage of supermarket aisles (see the following chapter). But the drop in nutrients in these areas alone has been so drastic, and so constant, that many nutritionists are now saying that in order to be assured of a healthy "diet" all people must routinely take daily dietary supplements—in the form of multivitamin pills.

"Absolutely," biochemical nutritionist Dr. Aileen Burford Mason told *Globe and Mail* reporter Picard. "When I hear people say, 'you can get all the nutrients you need from food,' I ask them: where is there a shred of evidence that is true? They are in denial."[30] Dr. Walter Willett, chairman of the Harvard University School of Public Health, agreed, calling a daily multivitamin "a good, cheap insurance policy."[31]

Unfortunately, Dr. Willett's cheap insurance policy may not be as good as he believes. Over-the-counter multivitamin and mineral tablets, of the kind we've all seen lining drugstore shelves in a bewildering variety of colorful and confusing packages, contain substances in pure, artificially concentrated form, as extracted via large-scale industrial processes. But the human body doesn't seem to work in a large-scale, industrial way. Whitney and Rolfes elaborate:

In general, the body absorbs nutrients best from foods in which the nutrients are diluted and dispersed among other substances that may facilitate their absorption. Taken in pure, concentrated form, nutrients are likely to interfere with one another's absorption or with the absorption of nutrients in foods eaten at the same time. Documentation of these effects is particularly extensive for minerals: Zinc hinders copper and calcium absorption, iron hinders zinc absorption, calcium hinders magnesium and iron absorption, and magnesium hinders the absorption of calcium and iron. Similarly, binding agents in supplements limit mineral absorption.

Although minerals provide the most familiar and best-documented examples, interference among vitamins is now being seen as supplement use increases. The vitamin A precursor beta carotene, long thought to be nontoxic, interferes with vitamin E metabolism when taken over the long term as a dietary supplement. Vitamin E, on the other hand, antagonizes vitamin K activity and so should not be used by people being treated for blood-clotting disorders. Consumers who want the benefits of optimal absorption of nutrients should use ordinary foods, selected for nutrient density and variety.[32]

The British government, in May 2003, went quite a bit further than the textbook authors, bluntly warning consumers in the United Kingdom that some over-the-counter vitamin and mineral supplements can actually endanger health, especially if taken in high doses. The British food standards agency, following a four-year safety review by independent scientific advisors, concluded that "long-term use of six substances, vitamin B_6, beta-carotene, nicotinic acid (niacin), zinc, manganese, and phosphorus, might also cause irreversible health damage."[33]

Sir John Krebs, chairman of the food agency, said "While in most cases you can get all the nutrients you need from a balanced diet, many people choose to take supplements. But taking some high-dose supplements over a long period could be harmful."[34]

A recent study published in the British medical journal *The Lancet* looked closely at the effects of regularly taking vitamin E and beta-carotene pills as a supplement to prevent heart disease. "Vitamin E and beta-carotene pills are useless for warding off major heart problems, and beta-carotene, a source of vitamin A, may be harmful," an Associated Press (AP) summary of the study reported.[35]

Researchers at the Cleveland Clinic Foundation gleaned similar conclusions from analysis of the pooled results of 15 key studies involving nearly 220,000 people–far more than needed to be statistically sound. "The public health viewpoint would have to be that there's really nothing to support widespread use of these vitamins," said Dr. Ian Graham, of Trinity College, Ireland.[36]

According to the AP: "The researchers found that vitamin E did not reduce death from cardiovascular or any other cause and did not lower the incidence of strokes. Beta-carotene was linked with a 0.3 percent increase in the risk of cardiovascular death and a 0.4 percent increase in the risk of death from any cause."

Because the pills didn't help, however, does not mean that vitamin E or beta-carotene themselves are not helpful in preventing disease. It only means that commercially produced pills that contain these substances in concentrated form may not help. Said the AP:

> The idea that antioxidant vitamins might ward off heart trouble was plausible. Test tube studies indicated that antioxidants protect the heart's arteries by blocking the damaging effects of oxygen. The approach works in animals, and studies show that healthy people who eat vitamin-rich food seem to have less heart disease.
>
> However, experts say that perhaps antioxidants work when they are in food, but not when in pills.[37]

Gulping jarfuls of orange, pink, or blue artificially concentrated vitamin tablets in an effort to offset the increasing nutritional poverty of our corporate/commercial food supply may actually end up making things worse, not better.

"Whenever the diet is inadequate, the person should first attempt to improve it so as to obtain the needed nutrients from foods," say Whitney and Rolfes.[38]

Great advice, but how can we follow it if the foods available at our supermarkets have few or no nutrients? If the trend lines over the past 50 years continue to hold true, it would seem that our food supply system is heading inexorably toward a diet made up largely of "non-foods" that contain increasingly fewer measurable nutrients, except for the relatively dangerous ones of fat, salt, and sugar.

Twenty or more years from now, if these trends aren't halted, will the "food" offered commercially in chain stores be nothing more than an attractively colorful but inert, sweet- or salty-tasting physical solid we swallow to give ourselves the illusion of eating, while we try hopelessly to obtain our real nourishment by juggling a smorgasbord of pills? "Hey, Jack, come on over for Thanksgiving dinner, we're having roast pill, with non-gravy!"

Whatever the future holds, for the past 50 years the nutrients have been leaching out of nearly everything we eat, leaving a vacuum that commercially produced vitamin pills can't fill.

And, as the saying goes, "nature abhors a vacuum."

Something else is already filling it.

MARY WASHINGTON IS A PRETTY YOUNG WOMAN,
with wide, innocent-looking eyes and a quiet, disarming voice. You
wouldn't think she'd scare anybody, but judging from her story in the
Westsider, a Michigan newspaper, she sure made some folks nervous.

Covering the consumer beat, she'd been assigned to investigate a
new supermarket phenomenon. As she told her readers:

> *Going to the grocery to buy meat has probably been the same routine
> for years: Pick your favorite pork chops, chicken, steak, etc. and pre-
> pare it at home, as usual. But what if that routine could threaten
> your life?*
>
> *People with high blood pressure, heart problems, or allergies had
> better start reading the package meat labels closely, because there's
> something new in chain store food most consumers aren't aware of.*
>
> *I went to the meat section of my two local grocery stores, and did
> some shopping around. I found that meat, especially boneless pork
> chops and boneless chicken, now has the words 'seasoned,' or 'pre-
> seasoned' in almost-invisible small print on the label. This disturbed
> me, so I went to the store butcher at [the first store] and asked what
> seasoning is in this meat.*[1]

Adopting a deadpan tone to match her innocent demeanor, she continued:

> He looked at me as if I was crazy, and asked why I was so concerned. I told him my mother has high blood pressure and if she is unaware that seasoning is in this meat, it could make her ill. Ordinary salt is actually sodium chloride—and sodium is something people with hypertension (high blood pressure) should avoid.
>
> According to the Family Health and Medical Guide, "a reduction in sodium intake is particularly important for persons with hypertension." Why? "Hypertension causes injury to the blood vessels, making it easier for atherosclerosis to develop. It is also the major factor in the development of stroke. In addition, hypertension makes the heart work harder, resulting in an enlarged heart, poorer function and congestive heart failure. Hypertension can result in damage to the eyes and kidneys as well." And sodium makes hypertension worse.
>
> Not to mention the possibility of ... whatever other unnamed "seasoning" is in this meat posing complications for persons with allergies.
>
> I also told the man I was a reporter for the Westsider. This seemed to make him nervous. His name tag said "Ron," but he refused to give a last name. He said he didn't know anything about the so-called seasoned meat, because all he did was package it and set it out on the shelves.
>
> The store manager, Melvin, also gave no last name and refused to speak on the record.[2]

Mary was persistent, however, and wouldn't give up.

> I asked why a fourth of the packaged meat on the meat counter was labeled "seasoned?" He said it was because it "has flavor" in it. I asked whether sodium was the best choice for flavor, reminding him ... about people with hypertension.

Melvin said if people have special needs then they should pay more attention to the foods they eat and make better choices when choosing their groceries. He seemed anxious for me to go away.

At [her other neighborhood store], the butcher said flatly "no comment!" He was not wearing a name tag at all. The store manager said he was busy and to leave my phone number to call me later on the issue. But I told him I would wait. After an hour, he came and said he needed to make a couple of phone calls first to know the right answers to my questions. He never got back with me. I also called him and left a message, but he ignored it.[3]

Since reading Mary's story I've made some forays of my own, to Canadian supermarkets, and found the same sorts of labels, and the same reluctance of store employees to answer questions about them. Most often, they simply shrug and plead ignorance. In some stores most meat labels now carry seasoning "instructions," advising shoppers to "season with pepper and or spices (No Salt)."

Why should diners suddenly need to be told—after centuries of deciding this question for themselves—how to season their meat? And why no salt?

Determined to get an answer, I followed the story up the food sales chain, from store clerk, to meat manager, to wholesale meat salesman, to a customer service agent for a meat packer in Manitoba, who at first told me "we have a lot of moisture-enhanced products going out now, because it tends to make the pork less chewy, more tender. But there is some salt in the moisture that's going into the pork already.... It's like a brine solution." When I mentioned the possible effects of salt on people with hypertension, however, she changed her explanation and claimed the no salt label was there "just because salt can make pork tough. If you salt it, it will be too tough." This explanation seemed to have more fishiness than pork about it. I've been putting salt on my pork chops for more than 50 years, and never noticed it making them tough.

Store employees may be reluctant to talk to reporters, or even ordinary shoppers, and customer service people may not want to alarm consumers, but the Internet tells all. Click on the Google search engine and type in the terms "pre-seasoned meat," or "moisture-enhanced meat," and the real story comes clear. That's what I did.

A SPREADING TREND

A website called the Virtual Weber Bullet explained the phenomenon succinctly:

> *Enhanced meat is becoming more and more popular in the United States. This trend is well established with pork and poultry and is spreading to beef products....*
>
> *Enhanced meat can be defined as fresh, whole muscle meat that has been injected with a solution of water and other ingredients that may include salt, phosphates, antioxidants, and flavorings. Regular meat can be defined as fresh, whole muscle meat that has not been injected or marinated....*
>
> *The problem isn't so much enhanced meat as a concept, but that it's becoming more difficult to buy certain fresh meat products in their regular versions—and fresh, natural, conventional meat is what most barbeque enthusiasts are looking for. Fresh pork is the best example of this trend. In some supermarkets, most fresh pork products, including spare and loin back ribs, butts, picnics and loins, are available only as an enhanced pork product. The same cuts of meat are not offered in their regular versions.*[4]

You and I, in short, have no choice. The choice has been made for us.

Whether we like it or not, we are obliged to accept whatever the meat packer and supermarket chain have decided is the way our meat ought to be flavored. Matters of personal taste are ignored. Like clothes off a discount department store rack, "one size fits all," and we can take it or leave it. If you like your meat with less pepper or salt than the chain has decided you should have, too bad. Big Brother knows best. Learn to love it.

And if your doctor has put you on a low- or no-salt diet due to high blood pressure or risk of heart disease, no matter. The corporate chain store has decreed that you *must* eat salt–or stop eating meat altogether. And bugger your health.

Die, or become a vegetarian.

There are, of course, some merits to the idea of being vegetarian. The majority Hindu population of India have been vegetarian for centuries and seem none the worse for it. They get their protein from other sources. But, damn it all, I'd like such a thing to be *my* choice, not the decision of some crew of corporate suits sitting in a boardroom somewhere with their accountants and advertising men and making up my mind for me! I'll wager most Americans and Canadians, if they'd take the trouble to stop and think about it, would agree.

The fact is, I *like* pork chops. I like them flavored to my own taste, not to someone else's. And I don't want to be forced to stop eating them altogether if I should unexpectedly develop heart problems. And what about those other ingredients, "phosphates, antioxidants and flavorings"? What if I'm allergic to one of them? Will it even be listed on the package label to warn me? Under the current supermarket regime, not likely.

And why are meat packers and supermarket chains switching to "enhanced" meats? The Virtual Weber Bullet site gives a number of reasons, but the most persuasive (given the added expense and complication for the packer of installing injection heads, pressure controls, filters, flexible needle mounts, and separate shut-off controls for each injection needle used in the enhancement process) seems to be "increased profitability":

> By "adding value" to meat by enhancing it, meat producers can charge more for their products and achieve higher profits. Also, by solving the problems of color retention and purge, enhanced meat

facilitates the trend toward case-ready meat—meat that is butchered
and packaged at the meat packing plant so that it's ready for display
and sale in retail stores. Case-ready meat is more profitable for meat
producers and for retailers, and it represents the future of meat in
America—and the demise of your local butcher.[5]

Nor is meat the only food product in which salt might be an issue.
For decades, fast food retailers like McDonald's have served french
fries with a liberal dose of sodium chloride. But recently, the international chain has adopted a new stance, agreeing to remove large quantities of salt from its products in Britain, including from the oil its fries
are prepared in.[6] Each serving of burgers, ketchup, or fries will have
up to 23 percent less salt than in the past.

WITCHES' BREW

Moisture-enhanced and pre-seasoned meat products may seem a high-
handed, profit-motivated intrusion on consumer independence, irri-
tating to anyone who values his or her gastronomical autonomy. But,
compared to what else is out there, they are relatively benign.

A mere pinch of salt, so to speak.

Today's mass-market foods contain far worse, with the general
rule being that the more highly processed the food product, the wider
the variety of hard-to-pronounce compounds inside it. In actuality,
what we are increasingly being forced to accept as "normal" fare
includes a witches' brew that would make Shakespeare's Weird
Sisters, cackling over their cauldron of "double, double toil and trouble," blanche.

The list of additives, pollutants, adulterants, and poisons is so
long, and due to so many different causes, that no single explanation
can cover them, or rank their importance. Perhaps the best method is
the one Hollywood uses to list the stars in a multi-star epic–alphabet-
ical order. Here are some of the "star" ingredients in the stuff that is
becoming our food.

ACRYLAMIDE

The organic chemical compound acrylamide, a derivative of acrylic acid (CH_2=CHCOOH) used industrially to make adhesives and textiles, wasn't thought of as a problem in food until April 2002. That year, Swedish scientists studying a group of tunnel workers who had been accidentally exposed to the compound on the job, revealed that they'd found high acrylamide levels not only in the red blood cells of the exposed workers, but also in people who hadn't been exposed.

The source for the latter group was traced to their food.[7]

Because acrylamide had been identified earlier by the World Health Organization (WHO) as a probable cancer-causing agent, the Swedish results caused an international stir, and a hunt for an explanation of how the chemical got into the food chain.

The hunt didn't take long. Acrylamide was found in varying levels in potato chips, French fries, crackers, breakfast cereals, and other processed foods whose manufacture includes heating, especially frying or baking. Scientists deduced they were formed by the amino acid asparagine and glucose sugar during the heating process.

Will this chemical "starlet" just breaking onto the scene prove a truly dangerous contaminant? The scientific jury is still out. Although acrylamide in high doses has been proven to cause genetic mutations in mice that lead to cancer, the level of the chemical in the average human's body at present is less than that in the lab mice. As the *Washington Post* reported:

> So far, officials say, they have not found acrylamide risks great enough to recommend that consumers avoid any groups of food or specific products. It remains uncertain whether people consume enough acrylamide in their food for it to be harmful, and whether the substance—which causes cancer in laboratory animals at high doses—is similarly hazardous to people, they said. But Terry C. Troxell of the FDA's Center for Food Safety and Applied Nutrition said yesterday, at a two-day advisory committee meeting on acry-

lamide, that the agency agreed with the WHO's conclusion that the discovery of acrylamide in many foods is a major concern and needs to be aggressively researched....

Troxell and other speakers stressed that ... its presence must be treated seriously.[8]

ADDITIVES (COMMON)

Unlike acrylamide, which was not deliberately–or in most cases even knowingly–introduced in foods, there is a long list of chemicals in what we eat which have been put there on purpose. Most of these additives, though not all, are there legally. That is, the manufacturers who put them there are not breaking any laws when they do so. These compounds include antioxidants intended to prevent food from going rancid, chelating agents put in to prevent discoloration, emulsifiers to keep water and oils mixed together, thickening agents, and flavor enhancers. There are hundreds of them, far too numerous to mention here, and as food manufacturers continue to experiment with new processes, more are being added every year.

In the U.S., the federal Food and Drug Administration is charged with regulating additives and assuring they are not dangerous. The laws and regulations the FDA goes by, however, are full of loopholes, as are the testing processes supposed to assess safety. For example, any additives that were considered safe by the FDA or USDA before the Food, Drug and Cosmetic Act of 1938 was amended in 1958 are exempt from regulation.

That is, if scientists working with the facts and testing processes available in 1938—three years before Japan bombed Pearl Harbor—thought a substance safe, then it is deemed safe for all time, regardless of what present-day research may have to say about it. Among these exempt-from-examination items are sodium nitrite (see below) and potassium nitrite, which are used to preserve cold-cuts and lunch meats.

At its "Chemical Cuisine" website (www.cspinet.org/reports/chem-cuisine.htm), the Center for Science in the Public Interest (CSPI) has posted a list of 73 common additives, rating them according to safety and describing their possible side effects. Anyone interested in the safety of the food they eat can download the list, print it out, and take it with them to the supermarket. You may have to squint a bit to read the fine print on some processed food labels, and not all labels on the shelf provide a comprehensive list of ingredients, but at least you can look up what is there and compare it with the CSPI's rankings.

For example, under *sodium nitrite, sodium nitrate*, you'll find this entry:

> *Meat processors love sodium nitrite because it stabilizes the red color in cured meat (without nitrite, hot dogs and bacon would look gray) and gives a characteristic flavor. Sodium nitrate is used in dry cured meat, because it slowly breaks down into nitrite. Adding nitrite to food can lead to the formation of small amounts of potent cancer-causing chemicals (nitrosamines), particularly in fried bacon.*
>
> *Several studies have linked consumption of cured meat and nitrite by children, pregnant women, and adults with various types of cancer. Although those studies have not yet proven that eating nitrite in bacon, sausage, and ham causes cancer in humans, pregnant women would be prudent to avoid those products.*
>
> *The meat industry justifies its use of nitrite and nitrate by claiming that it prevents the growth of bacteria that cause botulism poisoning. That's true, but freezing and refrigeration could also do that, and the USDA has developed a safe method of using lactic-acid-producing bacteria.*"[9]

The CSPI website also has pretty harsh words for most artificial food colorings and artificial flavorings, as well as for the sugar substitute aspartame, the flour "improver" potassium bromate, and various sulfites.

Among the various food colorings that have made it into the news media in recent years are Tartrazine (E102), Sunset Yellow (E110), and

Ponceau 4R (E124), dyes used to impart the typically orange-red hue to Indian dishes such as chicken tikka masala, served in Indian tandoori restaurants around the world. The three coloring agents, if taken over extended periods, are believed to be linked to hyperactivity in children, as well as to a list of other serious ailments, including asthma and cancer. As the British newspaper *The Guardian* noted:

> *Random tests ordered by Trading Standards officers in Surrey suggest 57 percent of Indian restaurants in the county use "illegal and potentially dangerous" levels of dyes to give the sauce its distinctive orange-red hue....*
>
> *Out of 102 curry houses sampled, only 44 were using the colorings within legal limits.*[10]

ANTIBIOTICS

Most of us think of antibiotics—biochemical substances produced by benign microorganisms that can inhibit or destroy harmful bacteria—as one of our best defenses against disease. After Alexander Fleming first isolated penicillin (produced by the mold *penicillium*) in 1928, and other scientists developed it for use as an antibacterial agent in 1941, the very word antibiotic became almost synonymous with "life saver."

Not anymore. Thanks in large part to the modern, corporate food industry, antibiotics are now on the list of dangers to human health. How this happened makes a kind of modern morality tale.

As most of us learned in high school biology class, bacteria are very small, incredibly numerous, and reproduce at a rate that makes rabbits look like they're practicing celibacy. In 24 hours, the offspring of a single *Escherichia coli* bacterium could outnumber the entire human population of the earth—and a certain number of that population of bugs will mutate.

All living things can mutate—experience a change in the character of one of their genes, or a change in the sequence of base pairs in a DNA molecule, which can then be passed on to their descendants. In large animals, like humans, mutations aren't all that frequent, but in a population of millions upon millions of bacteria, reproducing at whirlwind speed, there's "a whole lotta mutate'n goin' on," and each mutation can be passed on to millions of individuals within hours.

Antibiotics work by attacking bacteria in a variety of ways, such as breaking down cell walls or interfering with some vital step in the bacteria's metabolism. However, when an antibiotic attacks a bacterial colony, it doesn't always wipe that colony out. Some bacteria survive, either because they already had a genetic trait that blocked the antibiotic (called intrinsic resistance), or because they developed one while under attack (acquired resistance). These resistant bacteria can then go on reproducing, creating a resistant population. In general, the weaker the antibiotic attack, the more bacterial survivors there are—and the bigger the new, resistant population.

The best way to deliberately create a large, resistant population of harmful bacteria—if we are crazy enough to want to do this—would be to make many, many weak attacks on that species of bacteria with low doses of antibiotics. After each attack, there would be a fair portion of resistant survivors, and if the attacks are widespread enough, resistant bugs will soon be popping up everywhere.

This is exactly what our food production system is doing.

When a cow, pig or chicken "catches cold," that is, develops a mild bacterial infection, its milk or meat production goes slightly down while its body uses energy to fight the infection off naturally. In the old days of family farms with relatively small numbers of stock, nobody thought much of it. Only when a cow or sheep became significantly ill was medication used. But today's corporate factory-farm systems can't tolerate such minor blips. Maximizing profit is the name of the game, and nothing can be permitted to decrease production, not even a little bit.

Rather than wait for an animal to "catch cold," and suffer even a minor slowdown in milk production or weight gain, *preventive* doses of antibiotics are put into healthy animals' feed, as a sort of insurance against possible infection. The preventive doses, of course, are lower than those used to treat a full-blown, active infection. They are low, so-called "maintenance" doses—and their use has become more and more common, virtually guaranteeing that resistant bacteria strains will be popping up everywhere.

Of course, modern stock-raising methods aren't the only cause of the problem. Over-prescription of antibiotics by doctors treating human patients has also contributed to the development of drug resistance. But at least the doctors are treating actual sickness. The stockmen who feed perfectly healthy animals "growth-promoting" antibiotics are not.

As Michael Khoo of the Union of Concerned Scientists reported recently:

> *About 13 million pounds [of antibiotics] a year are fed to chickens, cows, and pigs to make them grow faster or to compensate for unsanitary conditions. That's about four times the amount used to treat sick people.*
>
> *Why is the use in animals a threat to public health? Because the overuse of drugs on factory farms creates antibiotic-resistant bacteria that are difficult to treat. These bacteria can make food-poisoning episodes last longer or recovery from surgery less certain. As bacteria become more resistant, people can no longer be sure that prescribed drugs will actually work.*[11]

The potential scale of the problem becomes clear when we look at some individual microbes. For example, the bacteria *Streptococcus pneumoniae* has become resistant to penicillin, and is

> *the most common cause of bacterial pneumonia (about 500,000 cases in the U.S. per year), is a major cause of bacterial meningitis (about 6,000 cases in the U.S. per year), causes about one-third of the cases of ear infection (about six million cases in the U.S. per year), and*

causes about 55,000 cases of bacteremia [bacteria in the blood, or "blood poisoning"] in the U.S. per year.[12]

The worst of the resistant bacteria strains are those that are immune to many different antibiotics, the so-called "superbugs." More than 90 percent of *Staphylococcus aureus* bacteria are now penicillin-resistant, and many of them are also resistant to methicillin, nafcillin, oxacillin, and cloxacillin, as well as other antibiotics.[13] *S. aureus* is the second most common cause of skin and wound infections, of bacteremia, and of lower respiratory infections. Some 40 percent of such infections are now due to multi-resistant strains. *S. aureus* blood poisoning "can be fatal within 12 hours."[14]

There are now also multi-resistant strains of *Salmonella*–a common cause of food poisoning,[15] and *Escherichia coli*, which is a major cause of diarrheal illness in children in the U.S. In severe cases of *E. coli* infection, dehydration can occur, "especially among children, in whom mortality may be quite high."[16]

More recently, scientists have reported a new strain of an ancient scourge: syphilis. This sexually transmitted disease, which can cause dementia, paralysis, and death, is caused by a microbe called *Treponema pallidum*, and until recently was easily cured by a few oral doses of the antibiotic azithromycin. The new strain is resistant to azithromycin, and is showing up in increasing numbers in syphilis patients. Incidence of syphilis itself has increased by more than 19 percent in the U.S. between the years 2000 and 2003.[17]

By constantly administering "sub-therapeutic" doses of antibiotics (that is, doses below the level needed to cure an actual infection) to the animals on farms, in feedlots, and in transport trucks carrying them to the slaughterhouses, meat producers create millions of resistant bacteria, with populations scattered all over the continent. Small residues of antibiotics may also end up in the meat sold in stores, which means we can be dosed with them when we eat the meat, leading to the creation of resistant bacteria in our own bodies.

So serious is the problem of drug-resistant bacteria that the European Union banned the use of growth-promoting antibiotics in meat and milk production in 1998. Such influential groups as the American Medical Association (AMA) and the World Health Organization (WHO) have called for major reductions in the use of such antibiotics in North America, but few producers have listened.

In fact, when the McDonald's fast food chain, responding to heavy consumer pressure, decided in June 2003 to ban meats produced with growth-promoting antibiotics, a storm of protest arose from the company's suppliers, some of whom claimed banning non-therapeutic antibiotic use would cause "a dramatic increase in animal disease"[18]— in other words, that not giving medicine to healthy animals would make them sick.

Yet continuing the practice may have contributed to what scientists call the "nightmare scenario," recently announced in the U.S. As the wire services reported in July 2002:

> Medical experts have long described it as the nightmare scenario of antibiotic resistance: the day when Staphylococcus aureus, cause of some of the most common and troublesome infections to inflict man, becomes resistant to the antibiotic arsenal's weapon of last resort, vancomycin.
>
> The nightmare scenario has arrived.
>
> The U.S. Centers for Disease Control has announced the first confirmed case of vancomycin-resistant staph aureus—known in the medical world as VRSA—found last month in a Michigan man.
>
> "The genie is out of the bottle," Dr. Donald Low, microbiologist-in-chief at Toronto's Mount Sinai Hospital says of the confirmation. "It's ominous."[19]

Low worried that the day is fast arriving when common infections like S. aureus won't be treatable with any antibiotics at all. That was the situation before penicillin was discovered. In those days, "many surgical procedures which now routinely save lives would have been too dangerous because of the risk of infection."[20]

Some scientists, seeing how slowly society is responding to the situation, believe most antibiotics will soon be useless, and research will have to turn to the relatively untested (in the West) use of bacteria-killing viruses called bacteriophages, or to chemical agents, as our only means of disease control—a high price to pay for a few more pounds of milk or meat, and a few more cents of financial profit.

There is also the possibility that antibiotic residues in food might cause allergic reactions in some people. A study presented to the European Respiratory Society's annual conference in 2003 reported that giving children an antibiotic before six months of age more than doubles the risk they will have asthma before their seventh birthdays. Babies who take antibiotics are also more likely to develop allergies to pets, ragweed, grass, and dust mites.[21]

Recently, the British Soil Association reported that people on diets involving high egg consumption may be in danger from lasalocid, an antibiotic commonly used by poultry farmers.[22] Residues of the drug were found in 12 percent of egg samples tested by the U.K. Veterinary Medicines Directorate. Although there are no reports of human illness induced by lasalocid, "similar drugs have been reported to cause severe illness, including paralysis and increased breathing and heart rates, and death in livestock such as cattle, turkeys and sheep. Lasalocid, commercially produced since 1977, has also accidentally poisoned dogs.

The U.S. Food and Drug Administration requires a specified period between the time of animal medication and time of slaughter, which is supposed to minimize the likelihood of such reactions, but the possibility remains.

ARSENIC

Well-known in its pure, inorganic form as a deadly poison (Lucrezia Borgia was alleged to carry it in a hollow ring, ready to tip into a victim's wine when he wasn't looking), the semimetallic element arsenic (As), in its less-toxic organic form, occurs naturally in food, water, and in the environment. The human body can tolerate a minimal amount of the organic variety, but exposure to too much of it over a long term has been associated with "cancer of the bladder, lungs, skin, kidney, nasal passages, liver, and prostate, according to the U.S. Environmental Protection Agency."[23] It has also been associated with "cardiovascular, pulmonary, immunologic, neurologic, and endocrine problems."[24]

Arsenic is a government-approved feed supplement used by poultry farmers to prevent parasite infections in chickens. The amount of arsenic found in young broiler chickens may be three to four times higher than that in other poultry, according to USDA researchers. How much arsenic residue in chicken meat is too much for a human to safely ingest?

The answer is, nobody really knows. But we're eating it.

BOVINE GROWTH HORMONE (BGH)

Bovine Growth Hormone (BGH, also known as recombinant bovine somatotropin, or rBST) is a hormone produced naturally in the pituitary glands of cows, which promotes both growth and milk production. Scientists can also genetically alter bacteria to produce BGH, permitting commercial laboratories to make massive, concentrated amounts of the substance, and sell it to farmers as a drug. Because BGH occurs naturally, and small residues have always been present in meat and milk, it hasn't been thought necessary to examine the effects of artificially produced or administered BGH, or to consider what might happen if larger-than-normal amounts should enter the food chain.

BGH is also denatured (viz., its function is changed) by the heat used in cooking meat or processing milk, and can be digested by the enzymes in the human gastrointestinal tract, giving scientists an additional reason to assume that it could have no effect if it enters our food supply.

However, a growing number of scientists and consumer activists believe such assumptions of safety are dangerously complacent, to the point of recklessness.

The most obvious reason for unease is the possible threat posed by BGH's effect on yet another hormone–called Insulin-like Growth Factor (IGF-I)—which is found in both cows and humans. IGF-I is extremely important because it appears to act as a sort of biochemical regulator or mediator that determines cellular response to various other growth hormones in various parts of the body. Abnormal increases or decreases in IGF-I can alter how the human body reacts not only to IGF-I itself, but how it reacts to the other hormones as well.

IGF-I is a potent "mitogen," or substance which stimulates cell division and boosts growth. Ominously, this can include not only the division and growth of normal, healthy cells, but also of cancer cells. As a survey of recent research on the subject by the Joint World Health Organization (WHO)/UN Food and Agriculture Organization (FAO) Expert Committee on Food Additives (JECFA) noted, Insulin-like Growth Factors are important mitogens in many types of malignancies.[25]

"Not surprisingly," the WHO/FAO authors add, "most of the cancers that IGF-I is associated with occur in tissues where IGF-I normally plays an important growth role, including the mammary, cardiovascular, respiratory and nervous systems, the skeleton and the intestinal tract."[26]

In other words, IGF-I may be a cause of potentially fatal cancers all over the body, but especially in the breast, colon, and smooth muscles. Says the report: "IGFs have been shown to be involved in breast cancer.... In the skeletal system, IGF-I has been associated with osteosarcoma (bone cancer). The tumor seems to strike children with the most rapidly growing bones.... IGF-I has also been implicated in lung cancer.... Five of eight human colorectal cancer cell lines were responsive to IGF-I."[27]

According to the report authors, "the weight of evidence indicates that rBST use [namely, BGH administered to cows] does increase IGF-

I levels in milk, substantially."[28] And the IGF-I in that milk, when ingested by humans "survives digestion."[29]

BGH's potential for increasing the presence of possibly cancer-causing IGF-I in milk and milk products has prompted the European Union to refuse approval for its use by European dairy farmers and to ban importation of dairy products from countries that use it. Canada has a similar policy. The U.S., under the Bush administration, has pursued an aggressive policy of threats and trade pressure to force the EU and Canada to accept BGH-laced products, but as of this writing the EU and Canada have resisted U.S. pressure.

The effects of BGH on IGF-I, however, are not the only reason why critics oppose its use. Cows that are regularly dosed with BGH also exhibit an increased susceptibility to one of the age-old plagues of dairy farmers: mastitis–a painful disease that affects cows' udders. The WHO/FAO report notes that in various trials of BGH-treated cows, mastitis incidence increased at rates varying from 50 to 76 percent, and that cases of rBST-associated mastitis "appear to be harder to treat than 'normal' mastitis."[30] That is, that to cure the affected cows, higher doses of antibiotics are required. Says the WHO/FAO report:

> Both increased incidence of mastitis and more severe or longer-lasting cases of mastitis can lead to greater antibiotic use. In the Vermont study ... there were more than seven times as many cases of mastitis in rBST-treated cows compared to controls, while the average length of antibiotic treatment was almost six times as long, leading to a 43-fold increase in the total duration of antibiotic treatment for rBST-treated cows, compared to controls. In the study of 15 commercial herds that found a 47 percent overall increase in mastitis in rBST-treated cows, antibiotic treatment doubled in rBST-treated cows compared to controls.[31]

In other words, a mastitis near-epidemic—prompted by greedy corporate dairy producers hoping to increase milk production and boost their accountants' bottom lines—requires those same dairymen to administer yet-heavier doses of the very antibiotics that could potentially create an alarming number of antibiotic-resistant bacterial strains, resistant not just to mastitis, but to a potential myriad human and bovine diseases (see Antibiotics, above).

At the same time, greater use of antibiotics in cattle is of concern "because of residues, which some authorities believe may cause adverse (i.e. allergenic) reactions."[32]

BROMATE AND BROMINATED DIPHENYL ETHERS (BDEs)

The chemical element bromine (Br) is found in a variety of compounds, including bromates (salts or esters of bromic acid, HbrO) and brominated diphenyl ethers (BDEs).

Early in 2004, a leading soft drink manufacturer was forced to withdraw its line of "pure" bottled drinking water from the British market when it was found to contain illegal amounts of bromate–which has been linked in studies with increased cancer risks.[33] More than 500,000 bottles had to be recalled, and not long afterward it was discovered that the bottled water had originally come from ordinary tap water, sourced from the company's factory in Kent.

Brominated diphenyl ethers (BDEs), commonly used as fire retardants in the foam used for furniture cushions, were reported turning up in rising proportions in the eggs of Great Lakes herring gulls in 2004, according to the Canadian Wildlife Service (CWS) of Environment Canada. Concentrations of the chemical, the CWS reported, had been doubling every three years since the early 1980s, and could be as dangerous as the highly toxic PCBs, which were banned in the 1970s. Said the Toronto *Globe and Mail*:

> The structure of brominated compounds closely resembles that of PCBs, prompting scientists to suspect that the two have similar biological effects. "There is no reason to believe that these things will be any different than the PCBs," said Ross Nystrom, an adjunct chem-

istry professor at Carleton University in Ottawa who has worked on
the research project. "They look the same. They've got the same kinds
of chemicals in them, and so far most of the research seems to be say-
ing they behave the same."[34]

In the past, birds contaminated with PCBs have had offspring with severe birth defects, "including extra limbs, malformed eyes and deformed beaks. Research on children suggests PCBs diminish intelligence."[35]

Are the BDEs detected in wild gulls likely to turn up in the eggs of domestic poultry, or at some other point in the food chain leading to humans?

No one knows.

DIOXIN

One of the legacies of the decades-long American attack on Vietnam was the effect of the Dow Chemical Company herbicide Agent Orange on the environment and people of Vietnam, and upon the U.S. servicemen and -women accidentally exposed to it. Visitors to Vietnam today can tour museum exhibits of the grossly malformed victims of the U.S. campaign to deny Vietcong guerrillas the cover of forests and farms by chemically destroying the tree and crop cover of the entire nation. Thousands of crippled children, and horridly malformed fetuses, were part of the effort to "destroy this place in order to save it" from what then-Secretary of Defense Robert McNamara later admitted was a tragically exaggerated, and mistaken, fear of "falling dominos."

Now the same substance that destroyed the lives and happiness of so many Vietnamese children and so many American veterans has turned up in the North American food supply—dioxin, known to biochemists as 2,3,7,8-tetrachlorodibenzo-p-dioxin (TCDD).

Medically, dioxin is described as: "a toxic, cancer-causing chemical. Initial exposure to this agent can produce chloracne [generalized acne], liver injury, and peripheral neuropathy [muscle weakness, impaired reflexes, and numbness, stinging and burning sensations].[36] It has been branded as a cause of severe birth defects. Long-term exposure to the substance can weaken the human immune system and cause cancerous tumors.[37]

North Americans, of course, are not being sprayed by the infamous Agent Orange, but they are subject to a never-ending barrage of dioxin, produced as a byproduct of numerous industrial manufacturing processes. Once released into the environment, dioxin tends to concentrate in certain food products, especially meat and other animal products such as eggs.

Canadian newspaper readers were shocked in 2002 by reports that dioxin—forbidden by law in any food sold in Canada—had been detected in foods imported from the U.S. in quantities up to 18 times higher than internationally accepted limits. "Eight out of 10 samples of pork, beef, and cheese contained chemical byproducts known as dioxins, even though the law stipulates that no such chemicals should be present in food sold in Canada," said the report.[38]

The World Health Organization (WHO) says dioxin levels in meat and animal byproducts must not exceed three parts per trillion for each gram of fat, if health and safety are to be maintained. But samples tested by the Canadian Food Inspection Agency (CFIA) found that a batch of eggs imported from the U.S. had dioxin levels of more than 90 parts per trillion—some 18 times higher than the limit set out by the WHO.

"By comparison, the dioxin content in a sample of Canadian eggs was 20 parts per trillion"[39] CFIA tested beef had dioxin levels of 23 parts per trillion, while cheese samples registered 12 parts per trillion.

Dioxin is also present in most fish caught by sport fishermen in North America, concentrating especially in the fatty tissues of salmon, trout, carp, and catfish. The danger posed is serious enough that some government departments, like the Ontario Ministry of the Environment, publish annual guides for fishermen, warning them away from badly polluted areas and estimating how much fish they can safely eat from various lakes.[40] Dioxin is also one of a mix of toxins found increasingly in farmed salmon, which recent studies have found are more contaminated than wild salmon.[41]

Only a few months after the alarming CFIA report, the U.S.-based Institute of Medicine (IOM) issued a warning to North American women, that they should "cut back their consumption of red meat, poultry and whole milk to reduce their exposure to dioxin."[42] According to this warning, dioxin "can build up in the body and, in their childbearing years, harm their babies."[43]

"Because the risks posed by the amount of dioxins found in foods have yet to be determined, we are recommending simple, prudent steps to reduce dioxin exposure while data are gathered that will clarify the risks," said Robert Lawrence, associate dean of the school of public health of Johns Hopkins University and chairman of the IOM committee that issued the warning.[44]

At the time of the advisory, the U.S. Environmental Protection Agency (EPA) reported that 150 kilograms of dioxins were released into the atmosphere in the U.S. in 2001, up from from 100 kg a year earlier.

In short, this deadly substance, which so devastated the population of Vietnam, is now present to such an extent in American and Canadian foods that women who plan to have children in future are advised to avoid eating meat, eggs, and dairy products altogether.

GENETIC MATERIAL (IN GENETICALLY MODIFIED, OR GM, FOODS)

When the first genetically modified (GM) food products began to appear on the market, a skeptical public immediately branded them "Frankenfoods," and reacted with suspicion. This reflex was ridiculed by the foods' corporate manufacturers as emotional and unscientific, but there is good cause to think that hesitation to accept the starry-eyed predictions of GM foods' most enthusiastic boosters may be no more than prudent common sense.

After nuclear power, the most controversial scientific advance of the twentieth century has been genetic engineering. Whole libraries have already been written on the subject, and as the twenty-first century began, it seemed the controversy was likely only to deepen.

The term genetic engineering refers to the manipulation of any genetic material, such as DNA (deoxyribonucleic acid), for practical uses. Normally, this involves introducing foreign genes into microorganisms so as to change their genetic code—and thus their basic nature.

First modeled in 1953 by British scientist Francis Crick and the American J. D. Watson, the "double helix" structure of the DNA molecule is the basic building block of the chromosomes contained in the nuclei of living cells, and carry within them the biochemical hereditary information–the codes–that determine the structure and function of most living creatures. Those codes consist of individual genes, segments of DNA that specify particular traits such as (in you and me) hair color, sex, body build, and so forth.

As anyone knows who has read Mary Shelley's nineteenth-century novel *Frankenstein, or the modern Prometheus*, or even seen the host of Hollywood productions based on it (my personal favorite is the 1931 Boris Karloff version), playing around with the basic stuff of life can get messy, resulting in things the players never expected, much less intended–and genes are, in every sense of the expression, the basic stuff of life.

Obviously, GM techniques hold enormous promise, particularly in medicine where they may eventually lead to cures for some of humankind's worst physical and mental afflictions. Applied in the field of agriculture, they hold just as much promise—and terrible dangers. The latter come under two headings: 1) the effects of GM foods on human health and, 2) the effects of GM crops on their surrounding environments.

The plain truth is, no one yet knows what those effects might be. And it's equally plain that far too many of the corporate players, who see quick fortunes to be made in GM crops and their applications, are unconcerned with such questions. There is a rush, like the blind, pell-mell stampede of the nineteenth-century Gold Rush, to grab patents and exploit them to the maximum, fast-tracking GM varieties onto an unsuspecting market before anyone has time to evaluate them, and to reap the short-term, maximum financial benefit regardless of long-term consequences—the consequences you and I, and our children, will eventually pay.

The question is not whether GM varieties should be developed, but whether they should be developed responsibly, with adequate pre-testing and evaluation of their environmental, market, and social impacts, or whether they should be introduced willy-nilly, recklessly, with no view toward any future more distant than the next fiscal quarter and no consideration beyond the potential financial bonanzas of greedy corporate CEOs. As the infamous Enron scandal, and others similar to it, ought to have demonstrated beyond question, the first concern of our modern corporate elite is hardly the public good.

The potential impact of GM foods on human health is largely unknown, but we do have some early indicators. A national advertisement published in 1999 by the Turning Point Project, titled "Unlabeled, untested....and you're eating it," noted that "while there have been no tests so far conclusively establishing that genetically engineered foods are harmful to humans, the potential dangers are significant enough to mandate long-term independent testing of GE food products before release into supermarkets."[45]

The ad listed areas of concern:

> **Toxicity.** *According to some FDA scientists, the genetic engineering of food may bring "some undesirable effects such as increased levels of known naturally occurring toxicants, appearance of new, not previously identified toxicants, increased capability of concentrating toxic substances from the environment (e.g. pesticides or heavy metals), and undesirable alterations in the levels of nutrients." In other words, scientists from the FDA itself suspect that genetic engineering could make foods toxic.*
> **Allergic reaction.** *FDA scientists also warn that genetically engi-*

neered foods could "produce a new protein allergen" or "enhance the synthesis of existing plant food allergens." And a recent study in the New England Journal of Medicine showed that when a gene from a Brazil nut was engineered into soybeans, people allergic to nuts had serious reactions. Without labeling, people with certain food allergies will not be able to know if they might be harmed by the food they're eating.

Antibiotic resistance. Many GE foods are modified with antibiotic resistant genes; people who eat them may become more susceptible to bacterial infections. Commenting on this problem, the British Medical Association said that antibiotic resistance is "one of the major public health threats that will be faced in the twenty-first century."

Cancer. European scientists have also found that dairy products from animals treated with bovine growth hormone (rBGH) contain an insulin-like growth factor that may increase the risk of breast cancer, as well as prostate and colon cancer.

Immuno-suppression. Twenty-two leading scientists recently declared that animal test results linking genetically engineered foods to immuno-suppression are valid.[46]

Fears that antibiotic-resistant genes engineered into GE crops could transfer to the guts of humans who ate them, thus leading to the encouragement of antibiotic-resistant strains of human diseases from meningitis and tuberculosis to gonorrhoea, were realized in 2002 when a British experiment was completed. Although some critics charged that the experiment, conducted by the UK Food Standards Agency, was biased and actually designed to minimize the chances of transfer, DNA from GM soya was nevertheless found to have survived passage through the small bowel of test subjects.

Dr. Mae-Wan Ho, of the Institute of Science in Society, wrote: "Despite the severe limitations placed on detecting GM DNA, and an experimental design biased towards negative results, irrefutable positive evidence was nevertheless obtained.... The latest finding is the last piece of damning evidence that horizontal transfer of GM DNA can indeed happen, has already been happening, and cannot be controlled if GM crops continue to be released to the environment."[47]

Antibiotic-resistant genes are put into GM plants as tags or markers, so that genetic engineers can tell when they have successfully inserted new traits into a plant. It is possible to remove the marker genes before the plant is released, but critics claim these precautions are rarely taken. Marker genes used in GM foods include lactam antibiotics, from which ampicillin and amoxycillin medicines are used as a first defense against chest infections.[48]

At least as dangerous as the threat to human health posed by GM crops is their potential to wreak havoc on the physical environment wherever they are introduced. GM crops—essentially manmade mutants–have no counterparts in nature, and thus nature has had no opportunity to develop the kinds of natural controls and defenses that normally keep other species in balance with the life forms around them. No one can predict the effects of introducing such novel organisms into our surroundings, especially if the genetic traits they carry end up being transferred to wild or domestic relatives of the genetically modified plant species. For example, as the Environmental Defense Fund's Rebecca Goldburg reports:

> In a highly controversial decision, the USDA in December 1994 allowed Asgrow Seed Company to sell squash genetically engineered to resist two plant viruses. The engineered squash will undoubtedly transfer its two acquired virus-resistant genes to wild squash (Cucurbita pepo), which is native to the southern U.S., where it is an agricultural weed. If the virus-resistance genes spread, wild squash could become a hardier, more abundant weed.[49]

The strategy chosen by several large corporations—to genetically engineer herbicide resistance into certain crops, which can subsequently be dosed with heavily toxic chemicals designed to kill their botanical competitors, while themselves suffering no ill effects—can boomerang on the engineers. The traits from the GE, herbicide-resistant plant may transfer to wild relatives. The authors of the textbook, *Plants, Genes, and Agriculture*, warn of the consequences:

> *If several crop species are made tolerant to the same herbicide and these crop species are used sequentially in a rotation, then there is considerable danger that "volunteers" from the first crop will become weeds in the subsequent crop. This phenomenon already occurs on a small scale now, but will be greatly aggravated by herbicide-tolerant crops.... All crop plants have wild relatives somewhere on the earth, and a certain amount of gene flow between the crop and the wild relative can occur where two populations grow side by side.[50]*

Another example of the unlooked-for consequences of introducing gene-altered organisms into the environment was noted in 1999:

> *A popular new variety of corn plant that's been genetically modified to resist insect pests may also be taking a toll on the Monarch butterfly, new research suggests.*
>
> *The gene-altered corn, which exudes a poison fatal to corn-boring caterpillars, was introduced in 1996 and now accounts for more than one-quarter of the United States' corn crop—much of it in the path of the Monarch's annual migration.*
>
> *Pollen from the plants can blow onto nearby milkweed plants, the exclusive food upon which young Monarch larvae feed, and get eaten by the tiger-striped caterpillars.*
>
> *In laboratory studies conducted at Cornell University, the engineered pollen killed nearly half of those young before they transformed into the brilliant orange, black, and white butterflies well known throughout North America.*
>
> *Several scientists yesterday expressed concern that if the new study's results are correct, then Monarchs—which already face ecological pressures but have so far managed to hold their own—may*

soon find themselves on the endangered species list. Other butterflies may also be at risk.[51]

A world without butterflies, jammed with out-of-control, herbicide-resistant superweeds, is hardly the picture intended by proponents of GM crops, but it is a possibility if such crops are introduced irresponsibly. In the absence of adequate regulation, some variant of that picture is almost inevitable.

"MAD COW" DISEASE (BOVINE SPONGIFORM ENCEPHALOPATHY, OR BSE)

The perfect example of a self-inflicted industrial wound, so-called Mad Cow disease (bovine spongiform encephalopathy, or BSE) first made its unwanted presence known in England, where cattle raisers, like their counterparts in North America, had hit upon a novel way of cost cutting. Rather than simply feed cattle the traditional kinds of rations—silage and hay—they decided they could add some protein to their herds' diets by grinding up the carcasses of dead cows and mixing them into the feed. The possibility that such an unnatural thing as turning grass-eating creatures into meat-eating cannibals, devouring their own species, might have unlooked-for consequences never entered into greedy cattlemen's heads.

But it had profound consequences, which are still being felt around the world.

As it turns out, feeding cows to cows leads to the spread of BSE, "a fatal condition that affects the central nervous system of cattle."[52] The malady is likely spread by an infectious agent called a prion, largely found in the brain and spinal cord of a diseased animal. It bores holes into the animal's brain (hence the term spongiform) and is always fatal. Worse, "a similar disease, called Creutzfeldt-Jakob Disease (CID), develops in people who have eaten the beef from infected cows."[53] The ugly consequences for anyone unfortunate enough to have done this are as follows:

Rapidly progressive dementia ... accompanied by neurologic symptoms, such as myoclonic [spasmodic] jerking [of muscles], ataxia [defective muscular coordination], aphasia [impaired speech function], visual disturbances, and paralysis.[54]

"There is no treatment, and the disease is fatal."[55]

After more than 100 cases of CJD broke out in England and Europe in the early 1990s, and were traced to Mad Cow-infected British cattle, British beef was banned throughout most of the world. The English cattle industry suffered a near-fatal blow from which it took years to recover. In the wake of this disaster, the European Union banned the use of tissue from animals as an ingredient in animal feed. The sale of cow brains as human food was banned outright.

A similar ban was put in place in Canada and the U.S., but with a significant loophole. While cattle, sheep, goats, and deer can't be given feed that contains protein from animals like themselves, their own bodies can still be turned into feed for chickens, pigs, and pets. In the U.S., chicken and pigs can also be fed *back* to cattle, and bovine blood is still fed to calves. As a result, it is still possible for U.S. and Canadian cows to have consumed BSE-infected material.

A scathing report from the U.S. General Accounting Office in 2002 found that the U.S. Food and Drug Administration had done a poor job of enforcing laws concerning cattle feed, and that manufacturers continued to include the banned proteins (namely those from the brain or nervous system) in their cattle feed.[56]

Worse, in the U.S. cow brains are still sold to human customers, along with cuts of meat, such as T-bone steak, stripped directly from the animals' vertebrae which may contain bits of the spinal cord. U.S. meat packing plants also use high pressure water and air or scraping methods to remove bits of meat off a cow carcass, and the recovered bits are added to hot dogs and low-quality hamburger. "A USDA survey last year found that more than one-third of products that contained this type of meat also contained some central nervous system tissue."[57]

In 2003, a single case of BSE was detected in a cow in Alberta,

Canada, which subsequent investigation discovered may have come originally from the U.S.A. Immediately, countries around the world—including the U.S.—banned importation of Canadian beef, and the Canadian beef industry shut down overnight. The ban cost Canadians nearly $20 million per day in lost sales, and after several weeks threatened to devastate the Canadian beef industry.

The consumer crisis of confidence provoked by these discoveries prompted some Canadian cattlemen to talk of "the end of our industry."

Meanwhile, scientists with the federal Health Canada department, who had warned of the danger in 2001, were saying "I told you so." In a 2001 report, Dr. Margaret Haydon had noted that "This is a disease perpetuated and spread by human hands, because we're feeding animals back to animals." Demanded her colleague, Dr. Shiv Chopra: "Why are we taking the risk? It's such a simple thing. Don't feed it and the disease stops. It doesn't spread. It's as simple as that. Why wouldn't they listen?"[58]

As of this writing, very few in North America have listened, and the danger remains.

The threat has even begun to affect other parts of the food industry. In January 2004, KOMO TV in Seattle reported that "some Asian markets, which have already banned U.S. beef, are now holding up or refusing shipments of potato products, thinking beef tallow [used to pre-fry some potato products like french fries] might be unsafe because of the mad cow scare."[59]

Still more chilling was the recent report that a new disease, similar to BSE, may have appeared. According to an article in the Manchester *Guardian*:

> *Tests on a heifer that died after five to six days of weakness in its legs and progressive paralysis have failed to identify any known condition, including BSE. A viral infection that damaged the white matter in the cow's brain is thought responsible for the death more than two months ago.*

The animal was at first thought to have died from botulism, a condition that is potentially dangerous to people through infected milk and food and is responsible for similar symptoms.

But that test proved negative, as did checks for West Nile virus, a mosquito-borne fever that Britain has so far escaped, louping-ill, a tick-transmitted disease also found in sheep, and other known conditions.

A spokesman for the Veterinary laboratories Agency last night said: "The long-term risk to public health is not known."[60]

METALS (HEAVY)

The human body needs small quantities of some metals, such as iron and copper, to function. But others—especially the so-called "heavy metals"—can seriously interfere with body functions. They do this by displacing, or chemically "shoving aside" the beneficial metals, preventing them from doing their jobs. **Lead**, for instance, displaces iron in the blood, but unlike iron cannot carry oxygen. When it replaces iron, the oxygen supply to the blood is reduced.

Called heavy because their atomic weights (a measurement of mass) are relatively greater than other metals, these substances are among the most dangerous contaminants in food.

For example, **mercury** (atomic weight 200.59, compared to the "light" iron at AW 55.847) is the cause of the deadly Minimata disease, which devastated the city of Minimata, Japan, in the 1950s and 60s, killing scores of people and leaving others blind, deaf, paralyzed, or brain-damaged. The Japanese victims were poisoned by eating fish and shellfish from Minimata Bay, which had been contaminated by methylmercury, discharged into the water as waste from local manufacturing plants.

Not long after the Japanese disaster, outbreaks of the disease occurred in North America, especially among native Indian populations whose members ate fish from northern rivers where sawmills had discharged mercury as a waste product. The publicity surrounding these outbreaks led eventually to government regulation of industry and the gradual elimination of mercury in many manufacturing processes. By the late 1970s most industrial sources of mercury had been stopped, but the metal, lingering as a residue, is still considered a major contaminant in fish, especially in the Great Lakes region.

Perhaps lulled by efforts over the past 30 years to reduce emissions of mercury and lead (the introduction of no-lead gasoline was a major step in this direction), the general public has come to regard heavy metal pollution as a problem of the past. Unfortunately, this perception may be badly mistaken.

In May 2003, Environmental Defence Canada released a report, based on unpublished Health Canada data obtained through federal freedom-of-information law requests, that the North American public is chronically overexposed to lead and **cadmium** because high levels of these metals are contained in many foods. As the Toronto *Globe and Mail's* Martin Mittelstaedt wrote:

> According to the report, elevated levels of cadmium were found in foods normally considered essential for a balanced diet, such as vegetables and cereals, as well as in items with no reputation for wholesomeness, such as potato chips and french fries. The highest lead concentrations were in frozen TV dinners and fish burgers, but ranking high were raisins, muffins, peaches, ground beef, and wine.[61]

Cadmium is a suspected carcinogen that also causes kidney damage. Lead can seriously damage the brain, and even trace amounts "can diminish intelligence, especially among children."[62] According to the authors of *Understanding Nutrition*:

> Lead damages many body systems, particularly the vulnerable nervous system, kidneys, and bone marrow. It impairs such normal activities as growth by interfering with hormone activity. It interferes with tooth development and causes dental caries in rats, and it has been associated with an increased prevalence of dental caries in peo-

ple as well. In short, lead's interactions in the body have profound adverse effects. [63]

As for mercury contamination, it is far from a historical curiosity of backward ages. It is still with us, especially in fish. As *Grist* magazine's Ann Harding reports, it is present in commercially important fish, such as tuna (especially albacore tuna), and there is considerable debate regarding what constitutes a safe level of consumption for humans. [64]

Heavy metals continue to be a major source of contamination in foods.

MEAT MIXES

Many health-conscious people, concerned about consuming too many fats or ingesting harmful substances such as steroids, attempt to limit their intake of certain meats. For example, some people eat no red meats, while others consume only lean meats, fish, or poultry, or only eat cuts taken from certain parts of an animal's body. Other shoppers, such as Jews or Muslims, may avoid pork, and eat only kosher or halal meats, for religious reasons.

Unfortunately for those who are trying to voluntarily control what kinds of meats they eat, some in the modern corporate food industry have decided that their preferences are not important.

Most shoppers are already aware that meat products like sausage, hot dogs, and various lunch meats and cold cuts are no longer obtainable as pure beef, except at premium prices. Most of the medium- to low-priced processed meat products contain a mixture, usually proclaimed on package labels as "may contain chicken, pork, beef, or other byproducts." At least in such cases the labels warn buyers what they are getting, if they care to read the fine print.

Recently, however, meat packers have begun using "hydrolized pork proteins, extracted by chemical hydrolysis from old animals, or parts of animals [namely, cattle and pigs] which cannot be used for food, such as skin, hide, and bone, to bulk up chicken. When injected into chicken with water, these proteins make the flesh swell up and retain liquid, rather like cosmetic collagen implants."[65] According to various reports, the practice is already widespread in Europe, and there is no reason to believe it is not employed in North America. According to one newspaper:

> *The Dutch have developed multi-million-pound technologies to insert proteins and water into chicken they import from Brazil and Thailand. The industry grew up to exploit an EU tax loophole which exempted certain imports of salted meats from tariffs. Adding water to the meat dilutes the salt, making it palatable, and enables the processors not only to escape being taxed, but also to sell water for huge profits. In some cases, as much as 40 percent of the meat's weight is added water.*
>
> *While the practice is abhorrent to Hindus, who do not eat beef, and Muslims, who do not eat pork, it is only illegal if the meat is incorrectly labeled. The adulterated chicken is heavily marketed to ethnic minority restaurants and wholesalers. According to sources in the trade, it is also distributed to caterers, pubs and clubs, and used by manufacturers for processed meat products.*[66]

Said a U.K. government spokeswoman: "it is particularly dishonest that consumers who may have elected to eat chicken for religious, moral or safety reasons should instead be getting chicken with beef or pork."[67]

MILK PROTEIN CONCENTRATES (MPCs)
Milk Protein Concentrates (MPCs) and casein (a milk protein that is the principal component of cheese) have become controversial since the turn of the twenty-first century, when major U.S. dairy processing companies began to import them and use them in making various milk and cheese products. Neither MPCs (not to be confused with Whey Protein Concentrates, or WPCs) nor casein are currently manufactured in the U.S., largely for economic reasons.

Nor—and hereby lie the roots of the controversy—are MPCs offi-
cially recognized by the U.S. Food and Drug Administration (FDA) as
legally permissible food additives. Their use by U.S. manufacturers is,
technically at least, illegal. They have not been subjected to the FDA's
years-long process of certification as GRAS (generally recognized as
safe) food ingredients, nor do they appear to meet legal food labeling
requirements. There is little scientifically supported evidence that
they pose any physical danger to consumers, but, as their opponents
point out, there is also—absent GRAS status—no firm evidence that
they don't.

More to the point, for the continent's dairy farmers, they are taking away a large section of the market for whole milk, which hitherto
was used as the base to produce many of the products now made with
MPCs. Whole milk costs more than MPCs, and thus represents a higher overhead for manufacturers.

There are two types of MPC: 1) MPC 0404.90, a blend of dairy
ingredients containing 42 percent protein, the so-called "low end"
MPC, and 2) MPC 3501, which contains up to 80 percent protein, the
so-called "high-end" product. Generally speaking, products made
with the "high-end" MPC give consumers better nutritional value-for-
money than those made with the "low-end" MPC. Unfortunately,
products on display in the dairy aisle of supermarkets give no indication whether they are low- or high-end.

At this writing, the legal status of MPCs remains undecided.

MISCELLANEOUS CHEMICAL/DRUG ARRAYS (SOUPS)

Researchers worried about pollution and food contamination have
traditionally focused on one disease organism, chemical, or chemical
group at a time, trying to isolate its specific effects. But more recent
studies have begun to look at whole arrays or "soups"—mixtures of
many chemicals and pollutants which may be more dangerous when
blended together than any one of them is alone.

The picture that is emerging is starkly frightening.

For example, New York's Mt. Sinai Medical Center, the Environmental Working Group (EWG), and Commonweal collaborated on a study, made public in 2003, in which a group of healthy people were screened for 210 chemical and industrial residues in their bodies. Said the EWG's Bill Walker:

Researchers at two independent laboratories found an average of 91 industrial chemicals, pollutants, pesticides, and other chemicals in the blood and urine of nine volunteers. In the entire group, a total of 167 chemicals were found. Like most of us, the people tested do not work with chemicals on the job and they do not live near an industrial facility.

Of the 167 chemicals we found, 76 cause cancers in humans or animals. Ninety-four are toxic to the brain and nervous system. And 79 cause birth defects or abnormal development. The chemicals found here ... [are] the consequence of lifelong exposure to industrial chemicals that are used in thousands of consumer products and linger as contaminants in air, food, water, and soil.[68]

A much larger study, the U.S. Centers for Disease Control and Prevention's second National Report on Human Exposure to Environmental Chemicals, was released the same year and painted an equally grim picture. The CDC had screened blood and urine from 2,500 Americans for 116 chemicals. "This report is, by far, the most extensive assessment ever made of the exposure of the United States population to environmental chemicals," said Dr. David Fleming, deputy director of public health for the CDC.[69] The report documented that dozens of man-made chemicals were present in adults and children, "with children generally having higher levels."[70]

At a press conference to release the findings of these studies, the EWG, the San Francisco Medical Association, and other participants warned of an accumulating "body burden" of mixed chemicals that are building up in the population, with as-yet unknown consequences. They cited disturbing reports like one in the Netherlands that documented "associations between … exposure to certain organochlorine chemicals and gender-specific behavior in children. Boys with relatively high levels of PCB exposure were less likely to engage in play behaviors typical for boys; girls were more likely to engage in play behavior typical for boys."[71]

Said conference participant Michael Lerner, founder of Commonweal:

> *I wonder how many other American families are experiencing learning disabilities, cancer, asthma, Parkinson's disease, autism, immune disorders, birth defects and so forth? Every disease that I've mentioned, there is either data that demonstrates or that suggests that environmental factors may be contributing to the increase, and chemical exposures may be part of the picture. There is an epidemic of breast cancer and there is as epidemic of many chronic diseases in this country and the question is, what is the contribution of this body burden that we are all bearing?"*[72]

Contaminants in foods are part of this phenomenon, as are drugs— some, such as antibiotics, introduced by modern industrial farming and stock-raising methods—that taint our water supplies. A recent federal Environment Canada report, for example, found that Canadian rivers and streams are contaminated "with a range of pharmaceutical drugs that present unknown dangers to people and wildlife," including "painkillers, anti-inflammatories, and prescription drugs used to treat epilepsy and blood cholesterol."[73] Also found in water supplies were antibiotics, anti-depressants like Prozac, and even drugs common in birth-control pills.

How all of these varied chemicals and drugs are interacting with each other in the bodies of those exposed to them, and what effects this unholy soup might have in the short- or long-term future, is simply not known.

It is extremely unlikely, however, that those effects will prove beneficial.

NANOPARTICLES (BUCKYBALLS) AND ATOMICALLY MODIFIED ORGANISMS (AMOs)

"Nano" is a prefix meaning one thousand-millionth, as in nanosecond. With the advent of nanotechnology in the late 1980s, a new level of technological sophistication was reached. The creation of microscopic-sized products became a leading research avenue, and the science has grown apace ever since. According to Webster's online dictionary, nanotechnology is:

> Any fabrication technology in which objects are designed and built by the specification and placement of individual atoms or molecules, or where at least one dimension is on a scale of nanometers.[74]

In short, this is a technology that focuses on making really, *really* small things, often so small as not to be discernible to the human eye. Those past masters of technology, the Japanese, have concentrated on this area, making rapid advances—especially in the realm of computer components. Research has focused on developing micro robotic technologies that permit nano-sized robots, or "microbots," to be used in assembling other nano-sized items.

Since the building blocks at nano-level are individual atoms or molecules (groups of atoms combined), raw material for manufacturing can theoretically be taken from anywhere. As one website devoted to the subject puts it:

> Manufactured products are made from atoms. The properties of those products depend on how those atoms are arranged. If we rearrange the atoms in coal, we can make diamond. If we rearrange the atoms in sand (and add a few other trace elements) we can make computer chips. If we rearrange the atoms in dirt, water, and air, we can make potatoes.[75]

Theoretically, at least, we could build "replicators," akin to those fictional devices used to make meals for *Star Trek* crews, that could take atoms from any source—garbage, chunks of turf, old car parts— and turn them into whatever products we choose. The potential is enormous, and is already making cash-register-ringing sounds in the heads of many a corporate CEO. But like so many "miracles of science" of the past, there is a flip side. The "wonders" of nuclear power became the horrors of Hiroshima, Chernobyl, and Three Mile Island—and nanotechnology could bring us worse.

Take "buckyballs," or "fullerines," for example. These are engineered, soccer-ball-shaped molecules of carbon, which researchers hope to use for a variety of nano-applications in making drugs, cosmetics, and fuel cells. They are already being manufactured in quantity in Japan. Unfortunately, these "miracle molecules" are far from benign. A recent meeting of the American Chemical Society heard a report from Dr. Eva Oberdorster on the effects of buckyballs on fish and other underwater life:

> [Oberdorster] described what happened when she exposed nine large-mouth bass to water containing buckyballs at concentrations of 500 parts per billion. After only 48 hours, the researchers found "severe" damage to brain tissue in the form of "lipid peroxidation," a condition leading to the destruction of cell membranes, which has been linked, in humans, to illnesses such as Alzheimer's disease. Researchers also found chemical markers in the liver indicating inflammation, which suggested a full-body response to the buckyball exposure.[76]

Oberdorster also found that buckyball-tainted water was fatal to an aquatic crustacean called the water flea. Half the flea population in her test were dead within two days of exposure.

The trouble with nano particles is that, because they are so new and so few observations have been made of their possible side effects, we simply have *no idea* what their effects on the environment–including on humans–might be.

This has not stopped researchers in Thailand from launching a nano-biotechnology project to modify rice varieties, including the prized jasmine rice. Researchers use a particle beam to drill a nano-sized hole in a rice cell, permitting the insertion of a nitrogen atom. The nitrogen atom stimulates changes–manmade mutations–in the rice's DNA that cause new characteristics in the plant.[77] Scientists at Chaing Mai University used nanotechnology to change the color of the leaves and stems of the Khao Kam rice variety from purple to green.

Some 200 transnational food companies, including the best-known brand names, are reportedly investing in nanotechnology. How will the mutant plants or animals they create for the commercial market affect the environment around them when they are released into the world? How will foods created by nanotechnology, or containing nanoparticles, affect not only fish or crustaceans, but the human body and digestive system? What could happen if the mutant varieties mutate again after release, and escape the confines of the farm?

Again, no one knows. As one Health Canada researcher commented, "It scares me silly."

ORGANIC CONTAMINANTS (A.K.A. ROT, SHIT AND DISEASE GERMS)
In the 1970s, a major furor broke out in Canada's Quebec Province, known in the press as *le scandale de la charogne* (the carrion scandal). Investigators found that a number of meat packers in the province were putting spoiled meat into packages and selling it. Organized crime was reportedly involved, and the racket included shipping charogne in trucks to other provinces, especially Ontario, and perhaps to the U.S. as well.[78] The story stayed in the headlines for weeks, and figured later in the proceedings of government inquiries. Arrests were made, plants were shut down—and everyone thought the problem was solved.

It was thus with a decided sense of *déjà vu* that many Canadians, in 2003, once again found themselves reading headlines about spoiled meat. This time it was in Aylmer, Ontario, where a plant was suspected of shipping meat unsafe for human consumption.[79] The plant was shut down for two weeks and its products were recalled. The ink was hardly dry on the Aylmer headlines, however, when another story followed. This time it was from Kingston, Ontario, where a packing plant actually staffed by convicts from the minimum-security Pittsburgh Institution was shut down by the province for engaging in "questionable" meat inspection practices.[80] The plant, whose products were mainly used within the prison system to feed other inmates, had its license suspended pending investigation.

The incidents became an issue in the provincial elections, and were followed by a formal judicial review of food safety in Ontario. The conclusions of the review were published in a 559-page report in 2004, which noted that "food-borne illness remains a significant public health issue in Ontario."[81] As if to underline that statement, in August 2004 one of the largest mass recalls on record of possible *E.coli*-tainted ground beef was ordered by the Canadian Food Inspection Agency. At least seven provinces and six grocery chains were affected.[82]

Out west, in the coastal Canadian province of British Columbia, the public was even more deeply shocked when health officials reported that a serial-killing hog farmer may have mixed the flesh of his human victims with his pork. Provincial health officer Perry Kendall said he felt compelled for "ethical" reasons to come forward with the news.[83]

Any Americans tempted to think such things are uniquely Canadian have only to refer to the shocking revelations in the 1997 investigative book, *Spoiled: The Dangerous Truth about a Food Chain Gone Haywire*, by Nicols Fox.[84] The author recounts tale after tale of contaminated food in the U.S., and its devastating impacts on the life and health of those who eat it. More recently, the best-selling author of *Fast Food Nation*, Eric Schlosser, warned his readers:

A nationwide study published by the USDA in 1996 found that 7.5 percent of the ground beef samples taken at processing plants were contaminated with Salmonella, *11.7 percent were contaminated with* Listeria monocytogenes, *30 percent were contaminated with* Staphylococcus aureus, *and 53.3 percent were contaminated with* Clostridium perfringens. *All of these pathogens can make people sick; food poisoning caused by* Listeria *generally requires hospitalization and proves fatal in about one out of every five cases. In the USDA study 78.6 percent of the ground beef contained microbes that are spread primarily by fecal material. The medical literature on the causes of food poisoning is full of euphemisms and dry scientific terms: coliform levels, aerobic plate counts, sorbitol, MacConkey agar, and so on. Behind them lies a simple explanation for why eating a hamburger can now make you seriously ill:* **There is shit in the meat.**[85]

This is the blunt truth. Meat packers are making us, quite literally, eat shit (about which more will be said in a later chapter). Nor is meat the only source of danger. Around the time that Canadians were hearing of the mass ground beef recall, Americans were being warned that fresh produce is "the new frontier in food-borne disease prevention." An outbreak of salmonella food poisoning sickened more than 300 people in five states, and Roma tomatoes were suspected as the source of contamination.[86] According to the U.S. Centers for Disease Control and Prevention (CDC), some 76 million Americans contract food-borne illnesses each year. Of that number, some 325,000 are hospitalized and 5,000 die.[87]

PESTICIDES

Pesticides, as the suffix "–cide," as in "homicide," indicates, are substances deliberately designed to kill things. They have no other purpose. The original intention, of course, is to kill only "pests," which is to say creatures, usually insects, whose interests don't coincide with our own. Unfortunately, chemicals that kill bugs are very often deadly to people and animals as well, if not in the original dose meted out to the pest population, then in concentrations that accumulate over time in an unintended, non-target species, such as humans. Even when they are not lethal to such unintended targets, these substances may prove damaging, debilitating, or crippling.

In 2002, the U.S. Environmental Working Group (EWG) tested two groups of preschool children in Seattle to see whether eating organic food reduced their exposure to pesticides, such as those belonging to the organophosphorus group, that harm the brain and nervous system of growing organisms. The tests found that children who ate conventionally grown food had concentrations of pesticide residues "six to nine times higher" than those who ate organic foods. As the study's researchers noted, children exposed to high levels of organophosphorus pesticides are at high risk for bone and brain cancer, and for childhood leukemia.[88]

Another EWG study, titled "Forbidden Fruit," analyzed samples from U.S. Food and Drug Administration records and found that nearly half of the registered contaminates found in non-organic food samples were actually legally *banned* pesticides. "The 10 most contaminated foods were strawberries, bell peppers, spinach, cherries, peaches, cantaloupe, celery, apples, blackberries, and green peas."[89]

Thus, children and adults who ate conventional, non-organically raised foods, were at significantly higher risk of falling prey to deadly cancers and other diseases.

TRANS-FATTY ACIDS

Trans-fatty acids, or trans fats, are solid fats produced artificially by heating liquid vegetable oils in the presence of metal catalysts and hydrogen. They are produced commercially in large quantities to harden vegetable oils into shortening and margarine. According to the Harvard University Department of Nutrition:

> Food manufacturers also use partial hydrogenation of vegetable oil to destroy some fatty acids, such as linolenic acid, which tend to oxi-

*dize, causing fat to become rancid with time. The oils used to pro-
duce french fries and other fast food are usually this kind of partially hydrogenated oil, containing trans fats. Commercial baked goods
frequently include trans fats to protect against spoilage.*[90]

Trans fats have a number of advantages for food manufacturing
companies. They can be used to extend the shelf life of processed
foods, give potatoes and doughnuts more flavor, and make crackers
and cookies crispier and crunchier. They improve food's appearance
and make it less expensive.

Unfortunately for the humans who are consuming them in greater
and greater quantities, they also have a number of negative properties.
The Toronto *Globe and Mail* summarized these on its website in 2003:

*The bad news is that the product—trans fatty acids—also increases the
risk of heart disease. Trans fats have been found to raise the body's
level of low-density lipoproteins, which can clog the blood vessels
with fatty deposits, and to reduce the level of high-density lipoproteins, which would otherwise clear away those deposits. A 14-year
U.S. study of 80,000 female nurses between 34 and 59, who were
asked to report regularly on what they had been eating, compared
that information with the number who developed heart disease. One
conclusion was that trans fats were twice as bad for the cardiovascular system as saturated fats. Another was that with every two percent
increase in trans fats consumed, a person's relative risk of developing
coronary heart disease doubled...."*

*The industry has made particular use of trans fats since warn-
ings about the dangers of high-cholesterol diets of saturated fats
(found in palm and coconut oils) made consumers rightly wary of
the saturated fat in their diet. Companies have been marketing their
products as cholesterol-free or low in saturated fat, without having to
mention whether they are high in trans fats.*[91]

Trans fats have been linked not only to heart disease, but also to increased risk of diabetes and Alzheimer's disease. According to the Harvard University Department of Nutrition website, "approximately 30,000 premature coronary heart disease deaths annually could be attributable to consumption of trans fatty acids."[92]

As already noted, this list of "star" ingredients in our foods is neither complete nor exhaustive, but only a brief indication of some of the things "out there" that can become "in here" if we eat them. If ordinary supermarket shoppers, looking at the list, feel vulnerable, that feeling is very likely a sound, mentally healthy response to the available data.

It is also a normal response to ask *why* this situation should be allowed to exist at all.

AS THE PREVIOUS CHAPTERS HAVE SHOWN, the nutritional content of food in North America has been steadily declining for at least the past 50 years, and probably longer. Meanwhile, the amount of toxic, or potentially toxic contaminants in the same food has been rising in recent years.

It would be daunting, if not impossible, to try to plot the hundreds of foods, and hundreds of nutrients in them, mathematically on one graph. But if we could, these trends—declining nutrition and increasing toxicity—would form an X, and the point where the two trend lines intersect, the crux of that X, would be a point of no return, the point where food has minimal nutritional value and serves chiefly as a toxic poison—the point, literally, of the End of Food.

We are fast approaching such an intersection. It's likely we could reach it even before the worst effects of global warming—the other great environmental threat of this era—are felt.

Why? What is causing this?

In North America, scant attention has been paid to these trends, still less to their causes. A pitifully small number of independent researchers, mostly exponents of so-called "organic" farming, have made tentative forays. But for the most part, the subject has been ignored.

Not so in the rest of the world, where a host of scientific studies and reports point the finger clearly and obviously at the now-predominant system of corporate/industrial agriculture, known colloquially as "factory farming." As Scully and Mulder of televison's *X Files* might put it, "the truth is out there," that is, outside of North America's mainstream media and this continent's largely corporate-research-grant-sponsored scientific establishment. To find the truth, one must step beyond the borders of the U.S.A. or Canada, and beyond the blindered limits of industry-funded science.

It's also a good idea, before looking at the scientific evidence, to have at least a rough understanding of how the industrial farming system came to be, and what it replaced. Not many present-day North Americans—few of whom still have any direct ties to the rural world—know the story. It's a tale worth retelling, at least briefly.

UNSETTLING THE AMERICAS

At the time Henry David Thoreau was writing *Walden*, in the early 1850s, the vast majority of people in both the U.S. and Canada were engaged in farming.

By 1993, less than 2 percent were.[1]

Farmer and social critic Wendell Berry calls the time in between, particularly the period following the Second World War, "the unsettling of America,"[2] when rural community after rural community was depopulated in a mass exodus to the cities. It was partly a technology-driven exodus, partly an economic one, and partly the result of deliberate, if misguided, "social engineering" (about which more will be said in a later chapter).

The technological drivers included mechanization, and an array of intensive, "high-input" cultivation methods. This postwar, capital-intensive way of farming matured first in North America and Western Europe, where petroleum products needed to fuel heavy machinery were relatively cheap and where the chemical firms that manufactured the required fertilizer, pesticides, and herbicides were centered.

It was rapidly accepted as the norm by most university agriculture faculties and government extension services because it was an "instant hit," producing bumper crops and huge surpluses of corn, wheat, and other commodities. It was assumed as the basic model by the scientific plant breeders who, in the late 1950s and early 1960s, launched the so-called Green Revolution, which carried the industrial farming model around the world–in some cases imposing it as an aid condition on Third World governments.

For those unfamiliar with its story, the Green Revolution began as a publicly funded research effort aimed at applying science to the food problems of developing countries. The International Rice Research Institute (IRRI) in the Philippines and the International Center for Maize and Wheat Improvement (CIMMYT) in Mexico united the efforts of research scientists, foremost among them Norman Borlaug of the U.S., in developing new plant varieties that, if dosed regularly with powerful fertilizers and protected by deadly pesticides and herbicides, would produce high yields to meet the needs of rapidly rising populations.[3] Initial results were highly positive, but only if looked at in very restrictive terms, without considering more than a limited number of economic, social, and environmental factors.

Years later, Gordon Conway and Jules Pretty looked at some of those overlooked factors and summarized the situation:

> Industrial activity has always resulted in pollution. But agriculture, for most of its history, has been environmentally benign. Even when industrial technology began to have an impact in the eighteenth and nineteenth centuries, agriculture continued to rely on natural ecological processes. Crop residues were incorporated into the soil or fed to livestock, and the manure returned to the land in amounts that could be absorbed and utilized. The traditional mixed farm was a closed, stable, and sustainable ecological system, generating few external impacts.
>
> Since the Second World War this system has disintegrated. Farms in the industrialized countries have become larger and fewer

in number, highly mechanized and reliant on synthetic fertilizers and pesticides. They are now more specialized, so that crop and livestock enterprises are separated geographically. Crop residues and livestock excreta, which were once recycled, have become wastes whose disposal presents a continuing problem for the farmer. Straw is burnt because this is the cheapest and quickest method of disposal. Livestock are mostly reared indoors on silage on farms whose arable land is insufficient to take up the waste.[4]

Farming, which used to be a family affair, had been reorganized on the impersonal, mass-scale industrial model. Traditional systems, which were organic, local, labor-intensive, family-oriented, and based on the diverse "mixed farming" that saw several crops grown in rotation and various livestock raised on the same farm, slowly disappeared.

In contrast, industrial agriculture featured systems that are non-traditional, inorganic, international, capital-intensive, market-oriented, and based on uniformity. Food was no longer raised, but manufactured, like steel or auto parts, in highly specialized operations—outdoor factories.

On traditional farms, a variety of crops were once grown in rotation–corn one year, beans or alfalfa the next–continually renewing the soil nutrient content and, by continually changing the micro-environment in each field, blocking disease and the reproduction of insect pests. Now "continuous cropping" and "monoculture" dominate. That is, the same immediately profitable crop, such as hybrid corn, is grown in the same fields, year-in, year-out, with no rotation, constantly draining the soil of the nutrients other crops might have returned to it.

Corn, for example, is a "heavy nitrogen" crop, which demands large amounts of soil nitrogen to grow. After a year in corn, a field has had its nitrogen stores seriously depleted. Legumes, like beans or peas, "fix" nitrogen, concentrating it in their roots and from there making it once again available to the soil. Traditional farmers would follow their corn with a planting of some legume. On industrial farms, however, the goal is to maximize production of a single crop, year after year. Leaving a field for a year, or even a season, in some other, slightly less profitable crop is simply "not on." Instead, the lost nitrogen is replaced immediately by chemical fertilizers purchased from plants which manufacture their concentrated, inorganic products via a variety of capital- and natural gas and petroleum fuel-intensive techniques. The replenishing job a crop of beans might have once done, while returning a modest profit, is now done by a pile of very expensive sacks of inorganic chemicals. Sacks which, since time is money, the wage workers on the factory farm must apply as quickly as possible, and which are often discarded only half empty, in local creeks and drainage ditches, polluting local waterways.

To pay for this chemical renewal, the industrial farm must continue to plant only the crop which gives the highest immediate financial return. And chemical fertilizers aren't the only expense. The unnaturally rich feast presented to pests and disease organisms year after year by the monocrop system absolutely depends on pesticides and herbicides to survive. Where once the life cycle of a noxious insect might have been broken when a farmer planted a different crop in that field, now that insect is constantly encouraged to feast and reproduce in a virtual insect-heaven made up, acre after acre, year after year, of its favorite food! And of course, the only defense against overwhelming insect outbreaks (that is, until the advent of genetic manipulation) has been the use of increasingly deadly, increasingly expensive chemical pesticides.

The watchword of corporate/industrial farming, which bases itself on the economic theories of the so-called Chicago School of economists, is "efficiency." But it is efficiency with a remarkably short-sighted set of parameters. They go (to deliberately exaggerate for the sake of illustration) something like this:

If the soil and weather of a state like Iowa are more conducive to growing corn than that of, say, Kentucky, or any other state, then all corn in the United States ought to be produced in Iowa. Kentucky should get out of the corn business. So should all other states. And Iowa should get out of the business of producing any other crop than corn. If the conditions for producing wheat are better in Saskatchewan than in any other Canadian province, then only Saskatchewan should produce wheat, and no other province should try it. Nor should Saskatchewan waste its time trying to produce any crop other than wheat. Year-in, year-out, on every acre everywhere, Iowa should grow hybrid corn, corn, and corn, and Saskatchewan should grow wheat, wheat and wheat. And whatever it takes to maintain this unnatural system, in terms of expensive chemical "inputs," is justified. To corporate accountants, eyes on the bottom line of the ledger books for the current business quarter, this is "efficiency."

In the real world, it's lunacy, and even corporate farms can't carry things to quite that extreme. But they get pretty close to the accountants' model–as close as they dare.

And the actual contents of the foods we are being offered in our grocery stores are showing the results of this kind of thinking. The scientific proof, as already noted, is hard to find in North America, but not elsewhere.

WEIGHT OF EVIDENCE

A good starting point for those interested in finding that proof is a book published in English in 1994 by an agronomist and soil scientist at the Swiss Federal Institute of Technology in Zurich, Switzerland, Dr. Ahmad Mozafar. He gave it the bland, textbook-like title *Plant Vitamins: Agronomic, Physiological, and Nutritional Aspects*, and its 412 pages are hardly the kind of reading you'd take to the beach on a summer Saturday.

But we—that is to say, people who must eat and would like to stay healthy—owe this guy. He went to an immense amount of effort to compile what was in 1994 the most complete and exhaustive survey of scientific research into how modern farming methods are affecting the quality of what ends up on our dinner plates—a significant weight of evidence. Before his book, which cites and summarizes in English hundreds of studies, the research was scattered and spotty, hidden in the obscure scholarly journals of countries whose languages few in the U.S. or Canada can read. As Mozafar put it in the preface to his book:

> *A relatively large portion of the literature on the factors affecting plant vitamins has been published in non-English (e.g., German, Russian, etc.) journals and has thus gone mostly unnoticed by English-speaking scientists. It was thus decided to cite all of the literature that could be collected irrespective of "age" a) because there is no other literature available or b) in order to be as comprehensive as possible ... it was decided to include a sample of the data in the form of tables or graphs ... some of which have appeared in journals hard to find in most libraries around the world.*[5]

Mozafar's pioneering work was followed, in 1997, by an international conference in Boston, sponsored by the School of Nutrition Science and Policy at Tufts University, on the topic of "Agricultural Production and Nutrition." A number of groundbreaking papers were presented at this conference, including the report by Britain's Anne-Marie Mayer mentioned in Chapter Two, and others by scientists from Germany, Sweden, Lithuania, India, Ghana, and other countries. They were later published under the title: *Agricultural Production and Nutrition: Proceedings of an International Conference.*[6]

Another scientific conference, this one held in Toronto, Canada, in 2002, continued the focus on this subject, and also saw its contributions recorded in print, under the title *Toward Ecologically Sound Fertilization Strategies for Field Vegetable Production.*[7] Outside of these three major collections, only scattered individual studies are available in English, such as the work of Japanese scientist Dr. Joji Muramoto of the University of California at Santa Cruz, on nitrates in leafy vegetables (see below).

What does this research, so difficult to locate in North America, tell us? It tells us that, if we stick to the currently dominant farming methods of industrial agriculture, we are in trouble.

That is to say, it gives us a very broad nudge toward drawing such a conclusion. There are no absolute, totally definitive studies available anywhere that say—beyond all possible reservations in even the most exceptional circumstances imaginable—that the industrial farming system, in each and every instance, is ruining our food. Science doesn't work that way. It doesn't deal in absolutes.

Of course, many defenders of the indefensible would like us to think it does. The "researchers" who worked for the tobacco industry insisted for decades, in the teeth of countless independent studies, that "there is no link between smoking and cancer," or between smoking and heart disease, etc. They insisted that unless we could prove that precisely the 216,000th cigarette smoked by John Smith after 20 years of his addiction definitely and individually triggered the onset of lung cancer in his body, to the exclusion of any other possible cause, no matter how far fetched, there was no "scientific" proof of a link. As a public relations strategy, it worked for an amazingly long time, but in the end the tobacco pushers had to pay billions of dollars in damages in class action lawsuits. It's entirely possible that the junk food/obesity controversy now making news in North America,[8] or the nuclear industry's longstanding denial of the effects of low-level radiation on the human body,[9] could someday lead to similar results.

As for the possible effects of industrial farming, the picture is not so obvious. Many of the system's defenders are perfectly reputable, honest scientists (like their brilliant original model, Norman Borlaug) with years of accumulated positive data to bolster their side of the argument. They are light years away from the pseudo-scientific public relations campaigns of the tobacco companies. And their motives are often undeniably noble: to feed the hungry and eliminate disease. Whatever might be the schemes and plans of the food industry's CEOs, the scientists who back them up do not make easy villains.

And yet.

Are they aware of that elusive, hard-to-find, not-always-available-in-English evidence? Have they read the French, German, Russian, Bulgarian, Japanese, or Iranian literature? An agricultural study needn't have come from the University of Wisconsin, Texas A&M, the University of California at Davis, the U.K.'s Reading University, or Canada's Guelph University to be valid and reliable. The best scientists sometimes turn up in the least likely places.

Take, for example, some of the studies summarized by Mozafar. Several deal with the effects on plants of artificially produced, inorganic nitrogen fertilizers, compared with those of organic (namely, based on carbon compounds), naturally produced nitrogen sources.

For those who did not do well in high school chemistry class, nitrogen (chemical symbol N) is found naturally in gaseous form in the earth's atmosphere—in the air, of which it comprises 78 percent—and is one of the basic requirements for life of most living organisms. Some, like ourselves, take nitrogen in by breathing, while other creatures must obtain it by other means. Plants take it up through their roots, but not in its pure form. For most plants to absorb nitrogen, it must first be converted, with other chemicals, into a compound they can absorb. When legumes like beans or peas are planted, beneficial bacteria concentrated in their roots absorb nitrogen and convert it into a form plants can use. Animal manure also contains nitrogen which becomes available to plants when applied to and processed naturally by soils.

In 1884, European researchers discovered the theoretical basis of how to combine hydrogen with atmospheric nitrogen industrially, to form a compound called ammonia (NH_3)–the compound used today as the starting point for manufacturing most inorganic fertilizers. At pressures of more than 2,200 pounds per square inch and temperatures between 400 and 500 degrees C, the two elements are combined according to the reaction $3H_2 + N_2$ (high temp. & press.) $= 2NH_3$.[10] The resulting ammonia, which is about 82 percent nitrogen, liquifies when compressed, and can be readily absorbed in water at concentrations of up to 40 percent.

The *Western Fertilizer Handbook* adds: "Although the supply of nitrogen from the air is virtually infinite, sources of hydrogen are limited. In the United States, almost all modern ammonia production facilities use natural gas as the hydrogen source. One ton of ammonia requires about 33,000 cubic feet of natural gas to supply the hydrogen required. Alternative sources, such as naptha, a hydrogen-rich hydrocarbon refined from petroleum, are frequently used in foreign plants."[11]

This is, in short, a heavy industrial process, using massive amounts of nonrenewable resources, and so are the subsequent processes that employ ammonia as a component in making various fertilizers, ranging from single-nutrient fertilizers like ammonium nitrate to multi-nutrient ones that combine nitrogen with varying amounts of phosphorus and potassium.

Applied in concentrated amounts, the various inorganic, manufactured nitrogen fertilizers may react very differently with crops and soils than do organic nitrogen sources such as manures or legume crops. This can impact the nutrient content of plants grown with inorganics. And thereby hangs a good part of Dr. Mozafar's 412-page tale. In Chapter Five of his literature survey, he writes:

> *Numerous reports indicate that increased use of [mineral] nitrogen fertilizers decreases the ascorbic acid [vitamin C] content in two important vegetables, namely potatoes and tomatoes, the extent of which appears to be relatively large. Since potatoes and tomatoes are among the major sources of vitamin C in human nutrition, the long-term effects of heavy use of nitrogen fertilizers in producing these crops on the vitamin C supply of humans needs to be investigated.*
>
> *Nitrogen fertilizers may also reduce the ascorbic acid concentration in leafy vegetables such as different Brassica species (Brussels sprouts, cauliflower, and cabbage), chard, and spinach.*[12]

Increasing the amount of nitrogen fertilizer, as it happens, often increases yield (the volume of a given crop produced per acre or per hectare) as well as the physical size of the vegetables in question. This of course makes producers happy, since they're getting "more bang for their buck" in terms of sheer sales volume (as comedian David Letterman likes to say, "volume, volume, volume: it's the name of the game!"). But volume doesn't always go along with quality. Writes Mozafar: "Sorensen noted that applying 600 kg N/ha increased the weight of each cabbage head by almost threefold (535.6 vs. 1836.5g/head) but reduced the ascorbic acid content by about 34 percent (71.0 vs. 46.8 mg/100g in the 0 and 600 kgN/ha, respectively)."[13]

Bigger cabbages, but with much less vitamin C.

The Swiss researcher points to another, unexpected effect of heavy nitrogen applications:

> *Nitrogen fertilization and the kind of soil in which plants are grown may also affect the amount of ascorbic acid in the fruits and vegetables long after they have been harvested, i.e., during their storage or processing.... In apples, foliar application of urea [$(NH_2)2CO$, a solid nitrogen fertilizer which contains the highest concentration of nitrogen—46 percent] did not affect the vitamin C content in the fresh fruits. After harvest, however, fruits from urea-treated trees lost their vitamin C more rapidly than those from untreated trees.*[14]

Still more ominous, Mozafar cites studies that indicate "heavy use of nitrogen fertilizers on plants, along with decreasing their vitamin C content, may also increase the concentration of NO_3 [nitrate] in their edible parts, sometimes to potentially dangerous levels."[15] He adds, "heavy use of nitrogen fertilization appears to have a double negative effect on the nutritional quality of fruits and vegetables since it increases their nitrate and at the same time lowers their ascorbic acid content."[16] As you may recall from the preceding chapter, nitrate can break down into nitrite, which in high enough levels can be converted in the human body to potentially cancer-causing nitrosamines.

As for the other two most common ingredients of inorganic fertilizers, phosphorus and potassium, Mozafar's survey found that "Kanesiro et al. reported that tomatoes grown in soil contained the highest ascorbic acid when plants were grown with no added nitrogen and low phosphorus supply."[17] In contrast, "potassium fertilization has been found to increase the ascorbic acid content in many different plants," although "plants subjected to too little or too much [potassium] fertilizer may contain less ascorbic acid than those supplied with an optimum amount."[18]

In later chapters, Mozafar looks at the differences in results when one compares the application of mineral (inorganic) fertilizers with application of organic (natural, carbon-based) fertilizers. He notes that

> heavy or exclusive use of inorganic fertilizers, along with the use of various plant protection chemicals, are the hallmarks of conventional farming. The "alternative" methods, however, rely solely, or mainly, on the manure, composts, and compounds such as bone meals as sources of plant fertilizers, use less or no plant protection chemicals such as pesticides, herbicides and fungicides, and the products are distributed without the use of artificial preservatives or dyes.... Here we concentrate on the effect of conventional versus organic methods of farming on the content of plant vitamins.[19]

Citing a number of studies, he notes first that "use of organic fertilizers does not affect the concentration of most plant vitamins, with the exception of thiamine, which seems to be higher in the plants grown with organic instead of inorganic fertilizers. The number of studies conducted up to now is, however, relatively small and thus an unequivocal conclusion cannot be drawn."

In other words, while there are strong indications that heavy use of inorganic nitrogen or other fertilizers may decrease beneficial vitamin C and increase potentially harmful nitrates, application of organic-source nitrogen may not have any immediate effects on some crops, other than the positive ones of generally increasing yield and perhaps increasing thiamine (vitamin B_1)[20]. There are, however, some notable exceptions, such as "Schudel et al. [who] reported that spinach and stock beet (*Spinacia oleracea* L. var. circla) grown with organic fertilizers produced less yield but contained lower nitrate and higher ascorbic acid than those fertilized with inorganic fertilizers."[21]

Though not cited in Mozafar's survey, a later study in California produced similar results. Dr. Joji Muramoto compared the nitrate content of leafy vegetables from organic and inorganic farms in the state, and found that "conventionally grown spinach contained significantly higher levels of nitrate than organically grown samples."[22] Of the samples taken, 83 percent of the conventional, inorganically grown spinach samples exceeded the legal nitrate limits of the European Union.

Regardless of the possible effects on plant vitamins, "one of the most consistent differences between plants grown with organic versus inorganic fertilizers is the lower nitrate content of the former," Mozafar emphasizes.

And once again, the Swiss researcher calls attention not only to the immediate results on plants, measured at harvest, but also points to the longer, post-harvest effects of chemicals:

> *Vegetables grown with organic fertilizer seemed to retain more of their original ascorbic acid **after a period of storage**. Also, spinach retained more (lost less) of its original vitamin C and accumulated less nitrate during storage when grown with organic versus inorganic fertilizers.*[23]

MICRO- AND MACRONUTRIENTS

The familiar nitrogen, phophorus, and potassium (NPK) content of most synthetically produced fertilizers represent, it should be realized, only a small part of the total nutrients needed by plants to grow, and which are contained in their tissues at harvest. Botanists have identified a total of at least 60 chemical elements in plants, "including gold, silver, lead, mercury, arsenic, and uranium,"[24] but not all of them are essential (as Mr. Spock might put it) for plants to "live long and prosper." After years of experimentation, scientists have concluded that 17 elements are absolutely required "for normal plant growth and development."[25]

These are divided into eight so-called "micronutrients," required in small or trace amounts (equal to or less than 100 mg/kg of dry matter), and nine "macronutrients," required in large amounts (1,000 mg/kg of dry matter or more):

micronutrients:
    molybdenum (Mo)
    nickel (Ni)
    copper (Cu)
    zinc (Zn)
    manganese (Mn)
    boron (B)
    iron (Fe)
    chlorine (Cl)

macronutrients:
    sulfur (S)
    phosphorus (P)
    magnesium (Mg)
    calcium (Ca)
    potassium (K)
    nitrogen (N)
    oxygen (O)
    carbon (C)
    hydrogen (H)

Of course, as already noted, science is not a business of absolutes. It's entirely possible that, at some future date, botanists may revise the list, adding an eighteenth or nineteenth element. But for the moment, the number is 17. And the majority of commercially manufactured inorganic fertilizers contain either nitrogen alone (one element) or nitrogen, phosphorus, and potassium—the famous NPK trio.

Three out of 17 *essential* ingredients.

Commercial fertilizers are given numerical designations consisting of three numbers, referred to as their "grade," representing the weight percent of nitrogen (N), phosphate (P_2O_5) and potash (K_2O) they contain. For example, anhydrous ammonia is graded as "82-0-0," that is, it contains 82 percent nitrogen and no phosphorus or potassium.

Where are plants expected to get the remaining 14 essential elements, other than NPK, if they are not contained in the manufactured commercial fertilizer, and the soil has been depleted of nutrients by continuous monocropping and the constant application of harsh pesticides and herbicides?

Hmmm. Well. Ahem....

Talk to an agronomist or soil scientist whose research grant funds come primarily from corporate sources, and you will be told that all of the other essential elements are so abundantly present in soils that fertilizer supplementation is simply not needed. Just keep dumping on the nitrogen, which brings higher and higher yields and bigger and bigger tomatoes, and things take care of themselves. Nature supplies the other elements, automatically.

Yeah. That's it.

But look at the food tables, mentioned in chapter two. Is that really it?

Mozafar has something to say on the topic. Cautioning that there have been relatively few studies published, he states that what evidence does exist indicates "increasing the supply of most other [micro and macro] nutrients seems to increase the concentration of ascorbic acid in a wide range of crops and under a wide range of experimental conditions."[26] He cites, for example, several experiments with calcium (Ca), which show that "an increased supply of Ca seems to increase the ascorbic acid content in several plants in all cases known to us."[27]

Not only do experiments show that increasing nutrients increases vitamin C in plants, but others show the opposite, namely that decreasing the nutrient supply has a negative effect on vitamin C:

> *Deficiencies of certain micronutrients are known to reduce the ascorbic acid content in some plants. In the Florida citrus fruits, for example, application of Zn, Mg, Mn, or Cu to sandy soils to correct their deficiency was noted to increase the ascorbic acid content in the fruits.*[28]

As for vitamins other than vitamin C, "based on the limited information gathered, it appears that the concentrations of most other vitamins studied are positively affected when increased amounts of various mineral nutrients are supplied to the plants."[29]

It seems only common sense, after all. If plants need 17 nutrients to thrive, giving them only one (N), or three (NPK) may not be the best strategy. It may increase the sheer size and volume of the crop, but simple logic would seem to indicate it might not be enough to make that crop a truly healthy, nutritious, vitamin- and mineral-rich food.

As noted in chapter one, however, in corporate farming production volume and physical appearance of the food on the store shelf (size matters!) are the main goals. Nutrition doesn't even make the list.

Papers presented at the 1997 Tufts University conference tended to support Mozafar's survey results. German scientist Joachim Raupp, for example, reported on an experiment conducted at the Institute for Biodynamic Research in Darmstadt spanning the unusually long period of 17 years, on the effects of organic versus inorganic fertilizers on yield and quality for several crops. "All vegetables had considerably lower nitrate contents with manure than with mineral fertilization: in carrots, an average of about 57 percent, in beets 74 percent, and in potatoes 60 percent as much nitrate as in MIN [inorganic, mineral fertilization] was found over four years," reported Raupp.[30]

He also noted the post-harvest effects of different fertilizers on the storage ability of plants:

> When potatoes, carrots and beets cut into halves or sliced were kept in plastic bags or preserving jars at room temperature for three to five weeks, clear differences occurred, in most cases in favor of the manure treatments. Vegetables grown with mineral fertilizer, in particular with the medium and high levels, became dark brown with rot, partly dissolved and slippery, or white with fungi. As a rule, those produced with manure looked at most only slightly changed and less covered by fungi. Abele (1987) documented these observations by impressive pictures.... Most of the MIN samples were very unappetizing.[31]

Researchers from Ghana, whose paper had focussed on soil types in Africa, summed things up well: "What people and animals eat determines to a large extent their health status. What the soil lacks in nutrients, the crops will also lack, as will, ultimately, human beings and animals."[32]

Similar conclusions were drawn by contributors to the 2002 horticultural congress in Toronto. In particular, Japanese scientists compared the results of organic and chemical fertilization of two popular oriental leafy vegetables.[33] Concluded the Japanese:

> By organic fertilization, quality of the leafy vegetables was improved, which was indicated by high concentrations of sugars and vitamin C, as well as low concentration of nitrate. The difference in quality between organic and chemical fertilized vegetables was more clearly indicated by the concentration ratio of vitamin C to nitrate.[34]

OTHER VARIABLES

Choice of fertilizer sources and their concentrations aren't the only things that can influence the end-quality of vegetables, in terms of texture, flavor, or nutrition. A host of other variables are involved, including three that most farmers and gardeners can't do much about, namely the basic soil type of their land, the long-term regional climate, and the short-term local weather where they farm. Only greenhouse growers are more or less immune to such influences.

Among the remaining variables that growers can control are the choice of crop variety to plant (touched on briefly in Chapter One, on tomatoes), the choice of exactly when to pick or harvest a crop, the methods used to harvest that crop, the kind of irrigation methods used (if any), and the choice of storage and processing methods. Each of these factors can make or break a crop, in terms of its nutritional quality, and any combination of them can react together, creating a variety of multiplier effects.

In North America, not much research has been done on any of these variables. But at least a few people have looked at the influence of harvesting methods on crop quality. And where crops like tomatoes were concerned, flavor or nutrition were never at issue.

In California, timing of the vegetable and fruit harvests and choice of plant varieties has depended in part on labor relations, or rather on the potential for labor exploitation. For decades, Mexican migrant farm workers, *los braceros* in Spanish, had been indispensable to California growers at harvest time. Whole families of Mexican men, women, and children, allowed entry to the U.S. on a strictly temporary basis, worked in the fields, performing backbreaking labor under the hot sun for a relative pittance. They had no health benefits, no pensions, and no defense against being let go at the grower's whim. Thanks to their availability, growers never had to bargain with American-born farm workers who would have demanded better wages and working conditions. As long as braceros were available for easy exploitation, growers geared their crop choices to varieties that responded well to hand-picking.[35]

Then came Cesar Chavez and his United Farm Workers Union. In the 1960s, the sweat-labor de facto serfs of the growers suddenly began fighting back, demanding fairer wages and better working conditions. The migrants had decided that it was time that the rules established for American workers as long ago as the 1930s and 40s began to be applied to them.

A battle royal ensued, during which some growers negotiated contracts with the UFW, improving workers' wages and conditions, while others remained adamantly opposed to the Mexicans' demands, insisting that they continue to accept their serf-like conditions without complaint.

Strikes and even national product boycotts became the order of the day. Pressure mounted on government to stop the exploitation, and U.S. labor unions, anxious to end the competition from cheap foreign labor for American-born workers' jobs, were part of the pressure. In December 1964, the bracero program itself was formally terminated by the U.S. government, cutting off the growers' source of cheap, exploitable Mexican labor.

In the midst of the fight, as it became apparent that the UFW was not going to back down, and that the choice was between paying workers a living wage and treating them fairly or having no workers at all to harvest the crop, growers began considering the latter option. Efforts to develop a mechanical harvester to replace the braceros were initiated A researcher at the University of California at Davis, G.C. Hanna, had anticipated the situation as early as the 1950s and had begun work trying to develop a thick-walled, hard and rubbery tomato variety (in short, a red tennis ball) that could withstand the rough handling of a mechanical harvester. By 1961 he had already developed at least two varieties that might do the job (VF 145A and VF145B). Later plant breeders would continue to develop tough, hard varieties, despite the fact that they had "somewhat fewer vitamins."[36] At around the same time, Hanna's University of California colleagues, with grower funding, had designed a successful mechanical tomato harvester.

An article posted on an anti-immigration website recounts the story with thinly disguised xenophobic relish:

My major at the University of California at Davis, power and machinery, brought me into contact with people developing harvesters for crops such as grapes, peaches, and tomatoes. The tomato project was particularly interesting. Several people contributed in various ways, such as developing a variety that could withstand mechanical handling. But the key element of the harvester proved elusive. This finally fell into place when Steven Sluka, a refugee from the 1956 Hungarian Revolution—some immigrants can be useful!— conceived the idea of cutting vines loose from the ground, lifting them, then shaking the tomatoes off the vines. His technique was the basis for the first successful mechanical tomato harvesters.

Growers in California were faced with the loss of workers who were hand-harvesting their crops. Politicians and labor had teamed up to discontinue the bracero program, so that wages paid domestic laborers could be driven up. However, UCD researchers, with grower funding, had just successively [sic] tested the mechanical tomato harvester. When braceros walked out of the fields, mechanical harvesters rolled in.[37]

The author of the article, whose final sentence is a lament that "rural America fills up with foreigners," is apparently unaware that the first Europeans to settle California (pushing aside its First Nations Indian inhabitants) were the Spanish, that the state was originally Mexican territory, and that non-Spanish Californians are thus the real "foreigners."

Today, 100 percent of the process tomatoes grown in California are mechanically harvested. The social consequences?

Prior to the introduction of the mechanized harvester in 1962 about 4,000 farmers produced tomatoes in California; nine years later only

600 of these growers were still in business. Before the new machine, 50,000 farm workers, mostly immigrant Mexican men, were employed as tomato pickers in California. They were replaced by 1,152 machines (each costing about $80,000).[38]

The search continues for tomato varieties that will permit a similar denoument in the fresh market tomato industry. The impediment here appears to be that machine harvesting of fresh market tomatoes "diminishes shelf life" because "damage from machine harvesting reduces storage potential."[39] A description of how a modern mechanical harvester works illustrates the problem:

[Machine harvesting is] based on a 'once-over' principle in which the entire plant is cut and carried over the harvester, where the fruit is then removed.... The harvester cuts the vine at or slightly below ground surface. The vines, together with any loose fruit that may have fallen to the ground, are gathered into the machine's feed conveyor by the counter rotation of the pickup disks and convoluted belts.... Fruit-laden vines, meanwhile, are transferred from the feed conveyor to a reciprocating mechanism that begins a shaking action, causing the fruit to separate from the vine. As the fruit separates, it is transferred to a conveyor located directly below the shaking section. From this lower conveyor, the fruit is routed and distributed onto sorting belts.[40]

Since process tomatoes are shipped relatively short distances, to plants where they are crushed, pureed, or liquified and made immediately into things like tomato paste, ketchup, or pasta sauce, storage potential and physical appearance aren't very important. So the damage caused by all that cutting, shaking, and bouncing around on conveyor belts is not as great a problem.

Due to the concentration of the fresh market industry in California and Florida, however, fresh tomatoes intended for direct sale in supermarkets—as whole, intact tomatoes rather than tomato products—must travel longer distances and last longer on the shelf than their process cousins, and still look good when they get there. Consumers "demand cosmetic perfection," reminds Mines,[41] and so hand picking is still required. But jostling around in trucks, over the hundreds of miles between the corporate field and distant supermarkets in states and provinces that no longer grow their own tomatoes, still requires tougher varieties.

Mechanical harvesting and long-distance shipping dictate variety choice in another way, as well. Under mechanical harvesting, "the grower cannot return any fruit unripe at the time of harvest. As a result, greater than 85 percent of the tomato field must ripen at the same time or the grower sacrifices a large part of his crop."[42] Thus only varieties that ripen uniformly, at the same time and same fruit size, and/or are thick enough to withstand long-distance truck transport, can be considered. Hundreds of vitamin-rich, tasty heirloom varieties that once may have been grown and hand-picked are no longer acceptable, and must be abandoned.

The overall message seems to be that, rather than pay a living wage to farm workers, whether Mexican or American, it's better for consumers to eventually get used to tough, flavorless tomatoes, with fewer vitamins, and in ever-diminishing variety, if the industry can manage to breed them.

And what's true for corporate tomato growers and their contract suppliers is true for mass-scale producers of most other fruit and vegetable crops. "Efficient" mass production, not flavor or nutritional value, decides which varieties survive.

RIPE OR UNRIPE?

The choice of exactly when to harvest a crop and how to ripen it can also affect the nutritional content of foods, especially of fruits and vegetables. And here again, industry needs rather than those of consumers appear to take precedence.

Because fruit harvested ripe from the vine could—after being transported hundreds or even thousands of miles across the continent, and then displayed on supermarket shelves for one or two days—very well be spoiled and rotten by the time a shopper sees it, industrial growers can't afford to wait to harvest fruit or vegetables when they are naturally ripe. In the "old days," when the fresh produce sold at local farmers' markets only had to be trucked a short distance from nearby growers' farms, and sold the same day, the problem didn't exist. But in a concentrated global industry with a global/continental reach, it does.

Corporate growers and their contract suppliers have solved the problem by harvesting fruit such as tomatoes at the so-called "mature green or breaker" stage, when the red color is just beginning to be faintly detected in the product. As Mozafar explains, the fruit are "then 'artificially' ripened in transit or at the destination market with the aid of ethylene" gas.[43] Often, rather than wait until this stage of the operations, plants are treated while still in the fields with the multi-use pesticide/plant growth regulator ethephon.

Some context is needed here.

Ethylene (C_2H_4) is an odorless, colorless gas produced naturally by plants during the ripening stage of their growth. If the gas produced by the ripening plants is trapped and kept in the atmosphere surrounding them, it multiplies the speed of the ripening process. Ethylene can also be produced artificially. As one of its commercial manufacturers explains:

> Ethylene, also known as the "death" or "ripening hormone," plays a regulatory role in many processes of plant growth, development, and eventually death. Fruits, vegetables, and flowers contain receptors which serve as bonding sites to absorb free atmospheric ethylene molecules. The common practice of placing a tomato, avocado, or banana in a paper bag to hasten ripening is an example of the action of ethylene on produce. Increased levels of ethylene contained within the bag, released by the produce itself, serves as a stimulant after

reabsorption to initiate the production of more ethylene. The overall
effect is to hasten ripening, aging, and eventually spoilage.[44]

Of course, in the mass-scale industrial food industry, nobody is
talking about using anything as primitive as paper bags to ripen fruit.
Commercial operations employ highly sophisticated, and highly
expensive, temperature-and-humidity-controlled "ripening rooms."
Monitored by pressure regulators and flow meters, these chambers
most often use ethylene generators, which produce the required gas
by heating a liquid chemical composed of ethanol (a form of alcohol)
and various catalysts.

While the small amounts of ethylene that are given off naturally
by plants are harmless, concentrated doses of it under industrial conditions can be far less benign. Mixtures of ethylene gas and air can
explode when the ethylene concentration exceeds 3.1 percent by volume. The gas, which has a suffocating, sweetish odor, is both an anes-
thetic and asphyxiant. "High vapor concentrations can cause rapid
loss of consciousness and perhaps death by asphyxiation," warn the
authors of a leading textbook on postharvest technology.[45] They add,
reassuringly, that "removal to fresh air usually results in prompt recov-
ery if the person is still breathing." Handled in liquid form, ethylene
can also cause burns to the skin or eyes. Given the risks and expense
involved in ripening room technology, some producers favor the use
of ethephon, in the field.

Ethephon, or 2-Chloroethylphosphoric acid, is a manufactured
chemical which, if mixed in a mildly acidic (above pH5) water solution,
spontaneously reacts with the water, releasing ethylene. Sold under a
variety of proprietary names, the chemical is sprayed on crops several
weeks before harvest, to maximize the percentage of colored fruit.

Ethephon is not exactly the world's most benign substance, either.
According to Cornell University's Extension Toxicology Network
website, the compound is "currently registered in the U.S. for use on
apples, barley, blackberries, bromeliads, cantaloupes, cherries, coffee,

cotton, cucumbers, grapes, guava, macadamia nuts, ornamentals, peppers, pineapples, rye, squash, sugarcane, tobacco, tomatoes, walnuts, wheat, etc."[46] Tests have determined that it can be "slightly toxic on a subacute dietary basis to bobwhite quail and mallard ducks," as well as to fish. Human subjects dosed with ethephon in controlled tests reported "sudden onset of diarrhea or an urgency of bowel movements, stomach cramps or gas and increased urgency or frequency of urination, and either an increase or decrease in appetite." Researchers also noted changes in cholinesterase activity in both animal and human test subjects' blood plasma.[47]

Whatever their more obvious drawbacks, the overriding virtue of ethephon and ethylene is that they allow growers to control ripening, which makes mechanical harvesting and long-distance transport practical.

What about their effects on the two factors most important to consumers, namely nutritional value and flavor?

Take tomatoes, once again, as an example. According to Mozafar's survey, multiple studies have shown that "vine-ripened fruits are higher in ascorbic acid [vitamin C] than those that are artificially ripened.... The reason for the relatively higher ascorbic acid in the vine versus artificially ripened tomatoes may be due to differences in the rate of ascorbic acid synthesis (or accumulation) in the fruits ripened differently."[48]

As for flavor, "tomatoes picked when fully or partly green or at the breaker stage and artificially ripened at 20 degrees C may be less sweet, more sour, have more off flavor, and less tomato-like flavor than those left on the vine to ripen."[49]

Artificial ripening strikes out on both counts.

WATER HAZARDS

Another lynchpin of industrial agriculture, particularly in states like California, is irrigation. "Across California, nine million acres of farmland rely on irrigation water ... with 80 percent of the state's developed water going to agriculture," notes David Carle in his guidebook,

Introduction to Water in California.[50] In terms of output, California is the number one agricultural state in the U.S., accounting for fully 55 percent of the nation's farm production, and it's probably safe to say that without intensive irrigation the state's agricultural industry would collapse.

Ironically, the very reasons why California has become the cornerstone of North American crop production—its soils and especially its climate—are also the reasons why irrigation is necessary. California has what climatologists call a "Mediterranean" climate, similar to that of Southern Italy or France's famous Riviera. Most of its rain, about 75 percent, comes in winter, while summers are hot and dry to the point where droughts are frequent. In southern California autumn brings the hot, dry Santa Ana winds and the grass fires for which the area is notorious.[51] Parts of the state are outright arid, such as the Imperial Valley and Mojave desert.

The overall warm climate allows crop production year-round, but in the hottest months irrigation is needed to maintain yields. The truly arid areas couldn't produce at all without it.

How does intensive irrigation affect the nutrient quality of crops?

The literature on this subject is sparse, but Mozafar's survey turned up enough evidence to indicate that it might affect it adversely, especially if the application of water is heavy. He cites numerous studies showing that rainy climate is known to decrease vitamin C in turnip greens, rose hips, onions, feijoa fruits, and black currants. He adds that "experiments conducted under controlled conditions have shown that increased water supply to the plants may reduce the ascorbic acid concentration in cabbage, cauliflower, celery, cucumber, muskmelon, radish, snap beans, and tomatoes.[52]

This does not mean, of course, that watering plants is harmful, only that there is an optimum amount of water beyond which a drop in nutrients may occur. Unfortunately, in order to maintain yields per acre and the physical size of various fruits and vegetables, commercial growers may have to go beyond that optimum. Mozafar sums it up:

Thus, if a plant vitamin happens to be reduced as a result of irrigation, then one is faced with a tradeoff between yield, taste, and vitamin content. For tomatoes, for example, the case seems to be relatively clear since dry growing conditions, which may reduce their yield, improve their flavor as well as their vitamin C content. Although growing tomatoes with less water may be something that home and hobby gardeners may want to consider, for a commercial farmer the economic consequences of lower tomato yield produced under water-limited conditions need to be taken into account.[53]

Another result of irrigation is the problem of soil salinity. Soils contain various kinds of mineral salts, a portion of which are dissolved in soil water. As they draw moisture from the ground, crops separate the water from the salts, leaving the latter concentrated in the soil. If a soil is already slightly saline to begin with, irrigating it and then sowing crops will tend to draw the salts up toward the surface, and concentrate them. If the climate in the area is hot, water evaporation will make the problem worse.

In the southern San Joaquin Valley in California, the soil was originally created from marine sediments that were naturally saline, and the area's hot weather increases water evaporation. As a result, salinity is a constant problem, especially in the western part of the valley. As Carle puts it:

To keep salination from increasing to the point where no plant can survive, farmers must apply more water than the plants need. The extra water flushes salts down into the groundwater below. In many places, soil overlies impermeable layers, so the excess water and salts cannot drain away. When they accumulate and reach back into the root zone, farming may become impossible. Around the world, including in the Fertile Crescent of the Middle East, salination has been the historic bane of irrigated agriculture.... The Westlands Water District [of the San Joaquin Valley] has some of the greatest salinity problems in California.[54]

According to Mozafar, "information on the effect of salinity on the vitamin content of plants is very limited." However, salinity "was shown to decrease the concentration of ascorbic acid in the leaves of peanut and cabbage, and in the fruits of tomato and okra."[55] In addition, sodium chloride (salt) "has been reported to reduce the concentration of carotene in the leaves of radish, cabbage, lettuce, and tomato."[56]

As far as large-scale, industrial crop production is concerned, the case is not definitive. But there is at least enough evidence available to conclude that the combination—concentrated inorganic fertilizers, heavy irrigation, harvesting plants before they are ripe, then ripening them artificially, selecting plant varieties primarily for toughness and cosmetic appearance, rather than nutrient value—may not be the best way to grow nourishing food.

When the question of the toxic residues from the herbicides and pesticides required for industrial production is factored in, the picture gets far darker.

PEDDLING POISON

The prevalence of various toxic chemical residues in our foods has been described in Chapter Three. Some of the hundreds of compounds mentioned come from sources other than agriculture itself–sources such as mining, manufacturing plants, automobile exhaust, chemical products ranging from hair spray to furniture polish, oil spills, and so on. But the actual producers of food are among the worst culprits, to the point where, in the 1980s, the U.S. Environmental Protection Agency (EPA) declared agriculture the largest nonpoint source of water pollution in North America.

On-farm, a wide array of pesticides, herbicides, and fungicides were or are used routinely, while still other compounds are employed during food storage or shipping. They range from the now-banned DDT and deadly organochlorides like dieldrin and aldrin, to the potent organophosphates and carbamates that succeeded them, to the somewhat more benign elemental sulphur used as a fungicide. As University of California horticulturist T. K. Hartz notes, "use of organophosphate and carbamate pesticides has been the foundation of pest control programs in most crops, including vegetables."[57] Author Cynthia Barstow emphasizes that "the EPA considers 60 percent of all herbicides, 90 percent of all fungicides and 30 percent of all insecticides carcinogenic: cancer-causing."[58]

Barstow adds that the use of conventional farm pesticides "increased from about 400 million pounds in the mid-1960s to nearly 850 million pounds around 1980, primarily because of widespread adoption of herbicides in crop production. Since that time usage has decreased somewhat, ranging from a low of 658 million pounds in 1987 to a high of 806 million pounds in 1996."[59]

In 1996, the U.S. Congress passed the Food Quality Protection Act, whose goal was to reassess some 9,600 pesticide tolerance acceptability standards and revise them. The process, as Hartz notes, "could result in the loss of vegetable crop registration for many organophosphate or carbamate pesticides."[60] But it is a slow, time-consuming job and, once done, won't eliminate the persistent residues of these poisons, which linger on in our environment years after being banned. DDT, for example, was severely restricted in the United States and Canada 30 years ago, yet continues to figure as a serious contaminant in present-day food sources, such as fish.[61]

Worse than the mere persistence of the residues of now-banned compounds in the North American environment is their presence in foods imported from other countries, where such compounds may not yet be banned. The peddlers of these chemical poisons have not stopped making them simply because they are now restricted or forbidden in North America or Western Europe. Ever-anxious to make a buck, at no matter what cost in human life or health, they continue to manufacture the same dangerous toxins, and export them to countries which have not yet passed laws to control them.

And with the current, corporate-sponsored global trade in foodstuffs, products contaminated with such substances are coming back to haunt North Americans. Barstow describes the actions of

> *some multinational chemical companies that continued to sell or "dump" large quantities of chemicals overseas because they could. They were legal over there, just not here. U.S. manufacturers export-ed more than 465 million pounds of pesticides in 1990, while more than 52 million pounds were banned, restricted, or unregistered for use in the united States. We won't even begin to discuss the ethics involved. Instead, we will simply look at what happens to the food sprayed with these known killers. We import it ... and eat it. This is referred to as "the boomerang effect," and it affects us in a big way.*[62]

She notes that, in 1986, "73 of 164 shipments [from overseas] con-taining illegal [in the U.S.] pesticide residues were allowed to reach the marketplace."[63]

Our food continues to be laced with toxins, banned or not banned, old or recently introduced, and will continue to be so for decades to come. That is, if we buy it from the international, corporate food industry.

MEATS AND PROCESSED FOODS

Fruits, vegetables, and grains are not the only foods affected by the corporate/industrial farm manufacturing process. As already noted in Chapter Three, everything from hot dogs to corn chips has suffered the effects of mass production technologies and the "logic" of the Chicago School economic "efficiency" mindset.

Take poultry as an example. In the now-receding days of the fam-ily farm, most small operations had a mixture of livestock. My own cousin, on whose Michigan dairy farm I spent some of the best sum-mer days of my youth, ran around 20 milk cows, mostly Holstein Frisian, and their milk was his main cash product. But he also kept chickens, which supplied the family with fresh eggs and broiler meat year-round. Occasionally, he raised the odd pig or two, and I seem to recall some ducks. Later, on my own small farm in Ontario, I kept a mixed flock of 20 layers, whose lives were fairly ordinary. During the day they ran free, either in a wire-enclosed chicken run or the open fields next to it, hunting for seeds, bugs, and greens in the grass and scratching in the dirt. There was room enough for them not only to walk and scratch, but even to fly short distances. At dusk they trooped back into the coop we'd set up for them in the barn, hopped up on their roost and slept, safe from foxes or other raiders. They laid eggs mostly in the coop, but sometimes in the grass—big, brown, thick-shelled country eggs with thick, bright yellow yolks. They were delicious.

And when the birds, after several years, got too old to lay eggs, they'd be killed for meat. This was done via a method now largely forgotten, which involves using a sharp implement, shaped something like a carpet knife, to sever a nerve on the inside roof of the birds' beaks. This paralyzed them and apparently anaesthetized them as well. A skilled chicken man could do a surprising number of birds this way, in a relatively short time. They'd lived full (for a chicken) lives, under fairly natural conditions. They died quickly and, at least as it looked to us, with minimal pain.

Compare this with the modern, corporate poultry operation, described in graphic detail by Karen Davis of United Poultry Concerns, Inc.:

The modern hen laying eggs ... is an anxious, frustrated, fear-ridden bird forced to spend 10 to 12 months squeezed inside a small wire cage with three to eight or nine other tormented hens amid tiers of identical cages in gloomy sheds holding 50,000 to 125,000 de-beaked, terrified, bewildered birds. By nature an energetic forager, she should be ranging by day, perching at night and enjoying cleansing dust baths with her flock mates—a need so strong that she pathetically executes "vacuum" dust bathing on the wire floor of her cage.

Caged for life without exercise while constantly drained of calci-

um to form egg shells, battery hens develop the severe osteoporosis of intensive confinement known as caged layer fatigue. Calcium depleted, millions of hens become paralyzed and die of hunger and thirst inches from their food and water.

In the twentieth century, the combined genetic, management and chemical manipulations of the small Leghorn hen have produced a bird capable of laying an abnormal number of large eggs–250 a year in contrast to one or two clutches of about a dozen per clutch laid by her wild relatives. The laying of the egg has been degraded by the battery system to a squalid discharge so humiliating that ethologist Konrad Lorenz compared it to humans forced to defecate in each others' presence. Researchers have described the futile efforts of caged hens to build nests and their frantic efforts to escape the cage by jumping at the bars right up to the laying of the egg.

Battery hens suffer from the reproductive maladies that afflict female birds deprived of exercise: masses and bits of eggs clog their oviducts which become inflamed and paralyzed; eggs are formed that are too big to be laid; uteruses "prolapse," pushing through the vagina of small birds forced to strain day after day to expel huge eggs. The battery cage has created an ugly new disease of laying hens called fatty liver hemorrhagic syndrome, characterized by an enlarged, fat, friable liver covered with blood clots, and pale combs and wattles covered with dandruff. In recent decades, hens' oviducts have become infested with salmonellae bacteria that enter the forming egg, causing food poisoning in consumers. Disease and suffering are innate features of the battery system....

Battery hens live in a poisoned atmosphere. Toxic ammonia rises from the decomposing uric acid in the manure pits beneath the cages, to cause ammonia-burned eyes and chronic respiratory disease in millions of hens. Studies of the effect of ammonia on eggs suggest that even in low concentrations significant quantities of ammonia can be absorbed into the egg. Hens to be used for another laying period are force molted to reduce the accumulated fat in the reproductive *to stop laying for a rve the hens for four f their body weight rugs such as chlor- can be part of this*

who must stretch onous mash in the ay her neck feath- ne mash particles bacteria causing the mash creating choice but to con-

blade once and old and again at ten grow back. researchers com- ... between the horn and bone of the beak is a thick layer of highly sensitive tissue. The hot blade cuts through this sensitive tissue, impairing the hen's ability to eat, drink, wipe her beak, and preen normally. Debeaking is done to offset the effects of the compulsive pecking that can afflict birds designed by nature to roam, scratch, and peck at the ground all day, not sit in prison; and to save feed costs and promote conversion of less food into more eggs, because debeaked birds have impaired grasping ability and are in pain and distress, therefore eating less, flinging their food less, and "wasting" less energy than intact birds. Diseases of Poultry states that "a different form of cannibalism is now being observed in beak-trimmed birds kept in cages. The area about the eyes is black and blue due to subcutaneous hemorrhage, wattles are dark and swollen with extravasated blood, and ear lobes are black and necrotic."

The battery system depends on debeaking and antiobiotics. Many of the antibiotics used to control the rampant viral and bacterial diseases of chickens in crowded confinement can also be used to manipulate egg production. For example, virginiamycin is said to increase feed conversion per egg laid, bacitracin to stimulate egg production, and oxytetracycline to improve eggshell quality. In Factory Farming, *Andrew Johnson says virtually 100 percent of laying hens in the United States are routinely dosed with antibiotics.*

At the end of the laying period, the hens are flung from the battery to the transport cages by their wings, legs, head, feet, or whatever is grabbed. Many bones are broken. Chicken "stuffers" are paid for speed, not gentleness. Half-naked from feather loss and terrified by a lifetime of abuse, hens in transit embody a state of fear so severe that many are paralyzed by the time they reach the slaughterhouse. At slaughter the hens are a mass of broken bones, oozing abscesses, bright red bruises, and internal hemorrhaging making them fit only for shredding into products that hide the true state of their flesh and their lives, such as the chicken soups and pies, school lunches, and other food programs developed by the egg industry to dump dead laying hens onto consumers in diced up form."[64]

The saddest thing about this description is not that it is true, but that it doesn't include the even worse horrors of how broiler chickens are raised for meat, or how young chicks are hatched and raised.

I still recall the surreal atmosphere of a corporate poultry shed I once visited with a group of fellow farm writers. Just before entering the massive building its proud manager warned us not to make noise or any sudden movement, so as to avoid what he referred to as "flips." We entered the shed, which was crammed with hundreds of thousands of peeping chicks in various stages of development, jammed into boxed-in squares on the concrete floor. The sound was absolutely deafening. The farm manager was telling us something, but all we could see was his lips moving while his words were blotted out by the monstrous shriek that resulted from all of those thousands of tiny but simultaneous peeps.

When we stepped back outside, I asked what flips were, and was told the word referred to chicks literally flipping over and dying of fright at any new or additional sight or sound. The tiny birds were so terrifically stressed by the bedlam they themselves created that the slightest addition—the straw that broke the baby chicks' back, so to speak—could kill them.

Anyone interested in reading the full story on modern corporate chicken production is invited—provided they have a strong stomach and maybe a bit of a masochistic streak—to read Karen Davis's book, *Prisoned Chickens Poisoned Eggs.* [65]

So far there is no conclusive evidence that the crowded conditions of modern factory poultry operations described by Davis have contributed to recent outbreaks of "avian flu" that have killed thousands of birds in the U.S. and Canada, sickened several human victims in both countries, and killed at least 20 people in Asia. Common sense, however, would seem to indicate that highly infectious disease organisms would be more likely to spread quickly in crowded conditions, where birds are also highly stressed and thus likely to be less resistant. And pumping the birds full of antibiotics–which kill bacteria but not viruses–would be of little preventive help.

According to the U.S. Centers for Disease Control and Prevention, avian influenza virus strains such as H5N1 have been associated with illness and death in humans, as have the subtypes H3N2, H2N2, H1N1, H1N2 and H7N3. [66] Scientists in China recently reported finding a virulent strain of avian influenza H5N1 in pigs, indicating that the virus is crossing species barriers in a particularly dangerous way. The World health Organization (WHO) warned that "if pigs were harboring both bird and human flu viruses, the two strains could interact, to create a strain capable of transferring easily to humans." [67]

So far, it seems that avian influenza, when it does affect humans, is a problem for farmers and poultry workers who are in direct contact with live, infected birds, rather than for consumers of poultry products processed for the market. But our frightening experiences with Mad Cow disease outbreaks ought to make us err on the side of caution.

Poultry, of course, is not the only meat product subject to the industrial production system. A similar tale emerges from Rick Dove's essay on pork production, "The American meat factory":

> *In industrialized hog factories, pigs are raised in intensive confinement for their entire lives in huge windowless structures, choked by their own foul stenches. Subject to disease from overcrowding and entirely deprived of exercise, sunlight, straw bedding, rooting opportunities, and social interactions that are fundamental to their health, factory hogs are kept healthy only by constant doses of subtherapeutic antibiotics, hormones, and toxic metals. Sows endure tiny crates that are too small for them to turn around, giving birth on bare metal grate floors, their babies taken away after only three weeks of nursing. Driven by frustration and depression, sows continually gnaw on the metal bars of their crates. Severe restrictions on the pigs' movement over a lifetime impede bone development, frequently resulting in broken legs. Injured pigs are "culled," sometimes by being dumped alive into waste lagoons. There are many accounts of brutal treatment of these animals, including teeth pulling, castration without anesthesia, and beating disabled sows unable or too terror-stricken to walk to slaughter. According to the U.S. Humane Society, one in five of all factory-raised pigs die prematurely, before reaching the slaughterhouse.*[68]

As for factory beef cattle—crowded in their thousands into feedlots where they may never see a green blade of grass, producing huge volumes of waste and pumped full of antibiotics—their lives are hardly less unnatural. One need only consider the cause of the disastrous Mad Cow disease outbreak: Beef cattle were being fed to other beef

cattle. What could be more unnatural than to turn herbivores into car-
nivores—and worse, into cannibals of their own kind?

As for the conditions under which beef cattle are gathered and
slaughtered, Gail E. Eisnitz detailed them only too well in her 1997
book *Slaughterhouse*. She interviewed two slaughterhouse workers:

> *Stationed near the blood pit, they'd toiled in constant fear for their
> lives. Juan Sanchez had quit after only a few days on the job, sure that
> he'd be crushed by a falling cow. Jose Alvaro, who had worked there
> for several months, described what it was like to work in the plant.*
>
> *Through the interpreter, Alvaro said, "My job was to wash the
> heads. I could see just about everything from where I worked." This
> included conscious cows thrashing while hanging from the rail, head
> skinners cutting spinal cords to stop the kicking, and a line speed too
> fast for the men to keep up....*
>
> *Alvaro said that even after workers were grazed by falling cattle,
> they were afraid to speak up. I asked him why. He snapped his fin-
> gers and pointed over his shoulder with a thumb, hitchhiker style.*
>
> *"Fired. Right away. On the spot," my translator told me. "Or
> moved to a different, much worse job, to get them to quit."*[69]

Another worker, Albert Cabrera, continued:

> *"In the morning, the big holdup was the calves. To get done with
> them faster, we'd put eight or nine of them in the knocking [killing]
> box at the same time. As soon as they start going in, you start shoot-
> ing, the calves are jumping, they're all piling up on top of each other.
> You don't know which ones got shot and which ones don't get shot at
> all, and you forget to do the bottom ones. They're hung anyway, and
> down the line they go, wriggling and yelling. The baby ones—two,
> three weeks old—I felt bad killing them so I just let them walk past.*
>
> *"But it wasn't just calves that went through conscious. It was a
> serious problem with the cows, and the bulls have even harder skulls.*

A lot I had to hit three or five times, 10 times before they'd go down.
There were plenty of times you'd have to make a big hole in their
head, and still they'd be alive.

"'I remember one bull with really long horns. I knocked it twice,"
he said. "Some solid white stuff came out—brains I guess—and it
went down, its face all bloody. I rolled it into the shackling area.
That bull must have felt the shackle going on its leg, it got up like
nothing ever happened to it, it didn't even wobble, and took off out
the back door, started running down Route 17 and just wouldn't
stop. They went out and shot it with a rifle, dragged it back with the
tractor...."

"What about the USDA people," I asked.

"They used to watch the animals stand up after I knocked them.
They'd complain, but they never did anything about it," he said.
"Never. The USDA vet would stand there to see how many live ones
were going in. I'd be shooting every one five, six times. She'd yell at
me, but she'd never stop the line. They didn't slow that line down for
nothing or nobody."[70]

Many of the cattle being killed in this inhumane operation were
physically sick, with what diseases the workers didn't know. As one
told Eisnitz:

"On bad days you'd have over 30 downers," he continued. "A lot of
them had fevers, some as high as a hundred and six. [The USDA doc-
tor] wouldn't let us kill those animals until their fevers dropped
below a hundred and five. She'd come down and take their tempera-
tures, but never did anything to help them. They'd lie out in that hot
sun, maybe for three days, before they died or [the doctor] told us we
could shoot them."[71]

And what are the results, in terms of human nutrition, of these
perverse, brutal, almost freakish practices?

Again, other than the general nutrient declines showing up in the food tables described in Chapter Two, the increases in such things as fat and sodium shown in the same tables, and the results of studies detecting the general presence of antibiotic residues and other contaminants in some meats, not much attention has been paid in North America to the actual links between production methods and meat quality.

What evidence is available, however, ought to give consumers and nutritionists pause.

Health writer Jo Robinson has researched the available scientific literature on the problem and published some of the data at her "Eat Wild" website (www.eatwild.com). She notes several studies showing that the meat of pasture-raised, grass-fed animals is far lower in fat than that of animals raised on grain in factory-style feedlots.[72]

Citing data from the *Journal of Animal Science*, she states that "a six-ounce steak from a grass-finished steer can have 100 fewer calories than a six-ounce steak from a grain-fed steer. If you eat a typical amount of beef (66.5 pounds a year), switching to lean grass-fed beef will save you 17,733 calories a year—without requiring any willpower or change in your eating habits."

Grass-fed beef is also apparently higher in vitamins than its factory feedlot counterpart:

> Meat from grass-fed animals is also higher in vitamin E. The graph below [from a study conducted at Colorado State University] shows vitamin E levels in meat from 1) feedlot cattle, 2) feedlot cattle given high doses of synthetic vitamin E (1,000 IU per day), and 3) cattle raised on fresh pasture with no added supplements. The meat from the pastured cattle is four times higher in vitamin E than the meat from the feedlot cattle and, interestingly, almost twice as high as the meat from the feedlot cattle given vitamin E supplements."[73]

Finally, Robinson notes that grass-fed meat products are higher in both the so-called "good fats," or Omega-3s, and in conjugated linoleic acid, or CLA.

People need a certain amount of fatty acids, and the human body can actually make some of them itself. The only ones it can't make are linoleic acid, referred to as Omega-6 (once thought of as "vitamin F"), and linolenic acid, called Omega-3 (the omega designations, for anyone who might be curious, refer to the locations of certain chemical bonds on the carbon chains that make up the acid molecules). These two compounds are described by nutritionists as "essential fatty acids," and the only way to get them is directly from the food we eat.

The reason they are considered good and essential is that the body uses them to maintain the structural parts of cell membranes, and to make important substances called eicosanoids. To quote the authors of *Understanding Nutrition*:

> *Eicosanoids are a diverse group of compounds that are sometimes described as "hormonelike."... Because eicosanoids help regulate blood pressure, blood clot formation, blood lipids, and the immune response, they play an important role in maintaining health. To make eicosanoids in sufficient quantities, the body needs the long-chain polyunsaturated omega fatty acids.*[74]

People who have ample amounts of omega-3s in their diets are 50 percent less likely to suffer a heart attack.[75] The omega fatty acids found in fish oils are also thought to be helpful in preventing macular degeneration, a common cause of sight loss in older people,[76] and some animal studies have suggested omega-3s may play a role in controlling some cancers.

Noting that "meat from grass-fed animals has two to four times more omega-3 fatty acids than meat from grain-fed animals," Robinson adds: "Omega-3s are formed in the chloroplasts of green leaves and algae. Sixty percent of the fatty acids in grass are omega-3s. When cattle are taken off omega-3 rich grass and shipped to a feedlot to be fattened on omega-3 poor grain, they begin losing their store of this beneficial fat. Every day that an animal spends in the feedlot, its supply of omega-3s is diminished."[77] She reproduces a graph from the *Journal of Animal Science* showing the steep decline.[78]

As for CLA, Robinson's site points to evidence from both animal and human studies that it "may be one of our most potent defenses against cancer." She notes that "in a Finnish study, women who had the highest levels of CLA in their diet had a 60 percent lower risk of breast cancer than those with the lowest levels."[79]

The Eatwild website, with the books and scientific journal studies it lists, is well worth a visit by anyone worried about assuring their family a healthy diet. So are health and environment-conscious magazines like the popular *Mother Earth News*, whose website recently featured an article on the decline of nutrients in our foods.

"American agribusiness is producing more food than ever before, but the evidence is building that the vitamins and minerals in that food are declining," wrote the authors. For example, "eggs from free-range hens contain up to 30 percent more vitamin E, 50 percent more folic acid, and 30 percent more vitamin B$_{12}$ than factory eggs."[80]

Would that Dr. Mozafar, could inspire one of his Swiss colleagues to produce the equivalent of his thorough survey of vegetable and fruit research–only this time covering the state of meat, poultry and dairy products. It ought to be an urgent priority for researchers in corporate-dominated North America as well.

But don't hold your breath waiting for it.

Instead, look for more corporate funding for research into so called "pharmafoods"—that is, foods deliberately designed by their manufacturers to create specific health or other physical or psychological effects in those who eat them. These include such things as a butter-like yellow fat spread containing phytosterols, designed to decrease the level of cholesterol in blood and thus reduce the risk of heart disease (recently approved for sale in the European Union), and teas containing mood-enhancing tryptophan.

Another such food is iodine-enriched chicken, for which a manufac-

turer asked EU approval, but was denied. The meat was created by feeding factory-raised chickens iodine-rich fodder, and was aimed at correct-
ing the diets of people with iodine deficiencies. Trouble was, no one
would have required a prescription to buy and eat it, and people who
did not have any iodine deficiency might risk serious health damage if
they ate it. When European authorities evaluated an application to market iodine-fortified eggs, they found that "the levels of iodine found in
the eggs were deemed to be so high that consumers easily could exceed
the safe upper level for the intake of iodine from these eggs.
Furthermore, iodine deficiency is not a general health problem."[81]

In other words prescription foods, put on the market willy-nilly
and without regulation, could become the equivalent of making powerful drugs—hitherto available only on a doctor's prescription—easily
obtainable on nothing more than the mere whim of every supermarket shopper. Today, potentially dangerous drugs, from anti-depressants and anti-psychotic medicines to drugs intended to reduce blood
pressure or improve blood clotting, cannot be dispensed without con-
sultation with and approval by a licensed medical practitioner who
understands their possible side-effects and knows the correct dosages.
Tomorrow, their equivalents may be freely available without prescription as "food" at the local supermarket, to ordinary people who have
no clue as to how or why they should be used.

The potential for confusion and outright abuse is enough to make
the hair on the back of your neck stand on end. Yet corporations, scent-
ing windfall profits, are going ahead full-tilt, accelerating research into
such products, with relatively little oversight by government.

BITTER FRUITS

"By their fruits shall you know them," Holy Scripture says, to which
wary shoppers might add "also by their vegetables, meat, dairy products, and pharmafoods." But the true nature of factory farming is not
only evident in the quality of what it puts on our plates. As the following chapter will show, its bitter harvest ranges much wider than that.

CALL HIM "BOLDUC."[1] He was, unfortunately, our neighbor on the Rang du Quarante (Fortieth Range) in the heart of Quebec's dairy country. Our farm was a small, part-time operation devoted to cash cropping hay and oats for local dairyfolk, most of them fourth- and even fifth-generation farmers whose awareness of the land, weather, and wildlife was so ingrained as to be almost a sixth sense. They loved the land and knew their place on it. Most milked herds of 20 or 30 Holstein-Friesians, sometimes brown-spotted Ayrshires ("Anglo cows"), on mixed farms of 100 or 200 acres, including orchards, sugar bushes, chickens and ducks, and the odd riding horse. They didn't get rich, but took care of their families, put the kids through school, and saved enough for retirement.

Bolduc was different: Like Caesar, he was ambitious.

When a neighbor whose children didn't want to farm retired, Bolduc bought the milk quota[2] and land, doubling his own holdings overnight. He installed tile drains in the new fields, built an open-concept feed barn where his cattle could be fed without resort to pasture grazing, and bought a new tractor—a big White "prairie-pounder," designed for use on Western Canadian wheat farms, that towered over him and could haul massive new machinery behind it. To make room for the tractor and machinery to turn around unimpeded, and to max-imize the area planted to crop, he razed the trees along the fencelines between his new and old fields, and filled in the ditches that divided them. Then he bought more land.

To pay for the new quota, barn, cattle, land, and machinery, he took out loans, on which the interest alone equalled the yearly profit of some of his smaller neighbors' operations. To meet the payments, he put all his fields to work growing high-yielding hybrid corn, a fast cash-return crop whose profits, with his milk sales, could be quickly funnelled into his creditors' pockets. He was running what govern-ment extension agents advised was an efficient farm, and hoped to make it still bigger and more efficient.

Our place bordered his land on two sides, and we soon found what the real meaning of "efficiency" was. The first sign was the half-empty fertilizer sacks, tossed into the stream between our prop-erties. Corn, especially modern, high-yielding hybrid varieties, is a hungry crop, drawing large amounts of nitrogen from the soil. As explained earlier, traditional farmers rotated the crops in their fields frequently, alternating corn with nitrogen-fixing plants like clover or beans so as to replace what the corn had taken. Rotating also pre-vented corn pest outbreaks, by eliminating the host plant for certain periods and disrupting insects' life cycles. But Boldduc couldn't afford that. Corn brought the highest fast-cash return, so he kept on planting it, year after year, in the same fields. To replace the lost nitrogen and other soil nutrients, he purchased expensive chemical fertilizers and spread them on his fields in ever-increasing amounts. To deal with weeds and pests, he bought equally expensive chemical herbicides and pesticides. Time was money, so the distribution of fertilizer, herbicides, and pesticides was done in a hurry, with large machines and temporary hired help, and the sacks were discarded carelessly.

By now, Bolduc's herd had tripled and was housed and fed almost entirely indoors. Unwanted manure from his huge barn, more than he could possibly use on his land, piled up outside. Rain liquefied it, carrying it away in runoff water.

There were fish in the stream when we bought our place, pumpkin seeds, sunfish, and a few suckers. The odd blue heron would hang around, looking for lunch. But after Bolduc's hired men began dumping used sacks there, and the manure pile began leaching into it, the fish died, and the herons stayed away. The water took on a turbid, greasy look.

What happened to the land was worse. The soil in the river valley we farmed was heavy blue clay, difficult to drain, and it took special care to maintain the soil's tilth and friability—its structure. The layer of good, nutrient-rich topsoil was thin. Under traditional farming methods, growth of soil organisms—bacteria, earthworms, and helpful insects—and retention of soil nutrients was encouraged. Cultivation was rarely overdone. But Bolduc's heavy machinery packed down the soil, creating a hardpan layer just below the depth of a plowblade. Overuse of chemicals killed off soil organisms, "burning" the powdery topsoil layer, which got thinner every year as the wind, now unbroken by fencerows, blew it away.

I walked onto my ambitious neighbor's property one day to look it over, and bent to touch the earth. It was no longer black, but gray, with the consistency of dry, fine gravel. What Bolduc was doing, I told a friend later, was not farming. It was outdoor hydroponics, the soilless growing method used in some greenhouses. After a few years of this treatment, the soil was dead, a mere physical solid for plant roots to grip to keep their stems upright, while all the nutrients were purely inorganic chemicals, washed through the field in solution. It was the most unnatural thing I'd ever seen, and couldn't possibly have continued many more years before the excessive costs would outrun Bolduc's ability to continually borrow and expand. And when it all collapsed, there would be nothing left to sell but a flat stretch of lifeless gray gravel, where once had been good farmland.

In the late 1970s we sold our own place and moved away, and so never had to witness the final scene.

Our ex-neighbor's story may have been an extreme case, at least in the speed with which it occurred. But it was far from exceptional in either its basic philosophy or the quality of its environmental effects. As Conway and Pretty, whose *Unwelcome Harvest: Agriculture and Pollution* was quoted in the previous chapter, made plain nearly 15 years ago, the currently dominant industrial farming system is an environmental disaster.

Its undesirable impacts can be grouped under six categories: pollution, soil degradation, wildlife habitat destruction, waste of freshwater resources, loss of biodiversity, and threats posed by introduced or "exotic" species (including those produced via the gene manipulation of bio-engineering).

POLLUTION

Conway and Pretty provide clear descriptions of the pollution problems created by industrial farming:

> *The primary environmental contaminants produced by agriculture are agrochemicals, in particular pesticides and fertilizers. These are deliberately introduced into the environment by farmers to protect crops and livestock and improve yields. Contamination is also caused, though, by the various wastes produced by agricultural processes, in much the same way as occurs in industry. The wastes comprise straw, silage effluent and livestock slurry, and, in the Third World, the wastes from on-farm processing of agricultural products such as oil palm and sugar. From the immediate environment of the farm contamination spreads to food and drinking water, to the soil, to surface and groundwaters and to the atmosphere, in some instances reaching as far as the stratosphere.*[3]

The worst offenders are the wide variety of herbicide and pesticide compounds, whose residues can enter the food chain, causing cancers and other diseases in livestock and humans, and impacting the natural environment in a myriad ways. For example, two of the best known herbicides, the compounds 2,4,5-T and 2,4-D, both contain the highly toxic contaminant dioxin (2,3,7,8-TCDD), already mentioned in Chapter Three, a chlorinated hydrocarbon created as a byproduct during manufacture.[4] One of the most deadly substances known, dioxin was present in the infamous Agent Orange herbicide used by American forces in Vietnam and, as noted earlier, is believed responsible by many researchers for thousands of abortions and birth defects among humans exposed to it.[5]

Herbicide and pesticide impacts on beneficial insects can be equally severe, as in the case of honeybees. According to Conway and Pretty:

> The first modern insecticides, the organochlorines, were relatively nontoxic to honeybees, but many newer organophosphates and carbamates are very hazardous. During the 1970s pesticides annually destroyed 40-70,000 bee colonies in California, some 10-15 percent of the total, while the annual national loss of colonies was estimated at half a million. At lower doses, pesticides may increase the aggressiveness of bees.
>
> Alfalfa is a crop particularly attractive to both wild and domestic honeybees. In one incident in Washington, large numbers of bees were killed by diazinon and two years later had only regained a quarter of their previous population level.[6]

When it is realized that honeybees are not only the source of a honey industry worth millions of dollars, but also necessary for pollination of farm crops from orchards to clover, the seriousness of such losses becomes evident.

Chemical toxins are also among the suspected culprits causing the surprising die-offs of North America's beloved Monarch butterfies, either by killing the butterflies themselves or denuding the landscape of their favorite food, wild milkweed.

Herbicide and pesticide contamination are particularly high where intensive, industrial farming is common: U.S. states like Iowa, Minnesota, and Ohio, where intensive corn and soybean cropping is practiced. It is also high in many Third World countries, where compounds like DDT, dieldrin, and aldrin, outlawed in northern countries, are often "dumped" and where both herbicides and pesticides may be applied without proper training or operator protection.[7] It is estimated that "about 40,000 people in the developing world die of pesticide poisoning every year."[8]

Industrial farming favors, even demands, use of toxic compounds, as the authors of *Unwelcome Harvest* explain:

> *In some parts of the world, pest problems first became serious with the expansion of irrigation and the increased use of chemical fertilizers. Year round irrigation makes double cropping possible and if the same crop is grown continuously, explosive pest outbreaks can occur. In general, the new cereal varieties tend to be more susceptible to pests and diseases and the loss of crop heterogeneity has favored high pest and disease populations.*
>
> *New practices, such as the direct sowing of rice, have led to increased populations of grass weeds, requiring more herbicides, while greater use of nitrogen [fertilizer] has heightened susceptibility to diseases. Often new practices come as packages of interlinked components. Direct seeding, for example, requires support from intensive herbicide use. This way, farmers can become locked-in to an intensive system where pesticides appear to be indispensable.*
>
> *Mechanization can also trigger pest problems. In the vegetable growing region of the Thames Valley, in the U.K., mechanization has produced economies of scale leading to large farms that concentrate on growing only three types of vegetable. There are fewer crop*

rotations and farms are essentially mono-cropped with higher-yield-
ing varieties susceptible to pests and diseases. As a result, pesticide
use has increased dramatically.[9]

Government subsidies aimed at boosting production, especially
for export, also tend to foster greater chemical use. In some Third
World countries, pesticide subsidies are as much as 89 percent of the
retail cost of compounds.

In addition to toxic agents intended to kill unwanted insects or
plants, inorganic chemical fertilizers—those mixes of nitrogen, phosphorus, and potasium (NPK) discussed in Chapter Four—pose pollution dangers when used excessively. Nitrates and phosphates not taken
up by crops enter the soil water, and from there the water supply of
animals and humans downstream from farms. When the nitrogen
level in drinking water (as nitrate plus nitrite) exceeds 10 mg/L it can
interfere with oxygen transport within the body. According to
Agriculture Canada:

> *children under one year old are particularly susceptible, as are cattle*
> *and young animals. There have been cases where excessive nitrate in*
> *the water of farm animals has reduced conception rates and*
> *decreased the number of live births.*[11]

Another unwanted side-effect of heavy fertilizer use is eutrophication of streams, ponds, rivers and other water bodies.
Eutrophication—oversupply of nutrients—can cause rapid, heavy
growth of algae and other plants, which ingest oxygen and give off
carbon dioxide. As a result, the underwater environment becomes
choked with plants and the oxygen balance is disturbed, leaving insufficient oxygen for fish and other creatures. Massive fish die-offs can
occur after such contamination. Dense growths of algae associated
with eutrophication can cause a further threat, from toxic compounds
secreted by such algal species as Prymnesium, and others. Fish and

livestock—particularly pigs—are vulnerable to various algal toxins.[12] Phosphorus stimulates the growth of blue-green algae, and agriculture is estimated to contribute approximately 40 to 60 percent of the phosphorus entering the Great Lakes through tributary rivers.[13]

In addition to these problems, serious difficulties are posed by creation of animal waste in industrial farming, mentioned briefly in the previous chapter. The size of the problem was first suggested 30 years ago by the American Association for the Advancement of Science (AAAS):

> As recently as 15 years ago, a one-man dairy operation included about 20 cows; today it has 50 to 60. Confined housing, well-engineered and executed layouts with good traffic patterns, automatic feeding and milking, and mechanical materials-handling equipment make it possible for a dairyman to enlarge his operation. There are farms with hundreds and even thousands of cows in production. The amount of wastes to be handled can vary anywhere from 100 to 200 pounds per cow per day. the dairyman from a 100-cow dairy operation will have to dispose of at least five tons of wastes per day during all the 365 days of the year (366 days in a leapyear), or something like 1,825 tons in a year's time.
>
> Nearly all beef animals spend one-third of their lifetime, three to four months, in feedlots, where they are fattened and polished before they are marketed. Waste disposal problems from the feedlots, even assuming they are well managed, arise from the large number of animals in each lot (anywhere from 500 to 5,000 head of cattle) and thus the large quantities of manure involved with no land to spread it on. [14]

Poultry farms cause even greater problems:

> In the major poultry producing regions, about 50 to 80 percent of the egg-laying hens and almost all broilers are raised under confinement. Practically all large poultry operations are highly mechanized, with

*automatic feeding and watering, optimum ventilation and tempera-
ture control, and automatic egg collection; but few have adequate
waste disposal systems. The number of birds on some poultry farms
exceeds one million.... The wastes from one million chickens would
be equivalent in strength to the wastes from a city of 68,000 people.
The managing and proper disposing of the wastes from a city of
68,000 people is a formidable task.... Several poultry operations in
several states, some with investments of more than $100,000, have
been forced out of business because of unsatisfactory systems for
waste management and disposal.*[15]

Manure represents a significant health threat to humans living downstream from farms. For example, the infectious bacteria salmonella can
survive for up to a year in liquid manure and is easily transmitted to people. According to *Agriculture Canada*, other infections transmitted to
humans via manure include: "anthrax, tularemia, brucellosis, erysipelas,
tuberculosis, tetanus and colibacillosis."[16] In the years since the AAAS
symposium, livestock operations have grown still more concentrated and
specialized, and disposal problems have increased accordingly.

The sheer physical noxiousness of these operations is at least as
bad as their possible health threat, as a recent news story about an
Oklahoma hog farm made clear:

*Julia Howell has stuffed feather pillows up her chimney. She said it
was to help keep neighboring hog-farm odors from seeping into her
home. She even caulked the windows in her two-storey, 17-room
home and bought an air cleaner, she said.*

*"It's not an odor, it's toxic fumes," she said. "They say it dissi-
pates in the air, but I disagree."*

*About three years ago, [a large firm] constructed a corporate hog
farm about a mile from the Howell home, she said. Nearly 10,000
hogs are housed at that facility, said Mark Campbell, vice-president
of operations for [the firm].*

A hog waste lagoon is less than a mile from her home, Howell said. "We have masks we wear outside, and sometimes I come in from working in my flower bed because of the toxic fumes," Howell said. "We have a golf driving range out here, and people used to come out here and use it and I used to give a few golf lessons, but that doesn't happen anymore."

Howell attributes the lack of interest in the driving range to the smell of hog manure.

"You can't breathe, it's that bad—regardless of what the hog people say," she said. "They say, well, you know this is agriculture, what do you expect? They tell us, if we're worried about it we're radicals against agricultural growth and economical development. And this is as far from the truth as it can be."

Howell and her husband, Bob, are longtime Hooker farmers. But they haven't invited dinner guests to their home for two years because of the smells from the hog farm.

"It changes your life," she said.[17]

There can be more than 100 kinds of substances in hog manure that cause odor problems. Most common are hydrogen sulphide (H_2S), a colorless gas that smells like rotten eggs, and ammonia (NH_3), but the others range up and down the chemical spectrum. As Bill Paton writes:

In a British study, 168 different chemical species were identified in manure slurry odor. Of these, 30 have odor detection thresholds lower than or equal to 0.001 mg/m3. Six of the 10 compounds with the lowest detection thresholds contained sulphur. Other researchers have identified a broad range of compounds in livestock manure including volatile organic acids, alcohols, aldehydes, amines, carbonyls, esters, sulfides, disulfides, mercaptans, and nitrogen heterocyclics.

Amine emissions have been studied because of their inherent toxicity and the potential carcinogenicity of their reaction products,

particularly the nitrosamines. Trimethylamine (TMA) is the pre-
dominant amine emitted from animal wastes with smaller odor con-
tributions from monomethylamine, isoproprylamine, ethylamine, n-
butylamine, and amylamine. Secondary amines, dimethylamine,
and diethylamine have also been identified inside livestock houses.

The organic portion of manure consists of proteins, carbohydrates and fats. Anaerobic decomposition degrades the proteins to
ammonia, hydrogen sulphide and short-chain organic acids. The carbohydrates decompose to organic acids, which are further converted
to alcohols, water, carbon dioxide, and hydrocarbons, especially
methane. The fats degrade to fatty acids and alcohols, and the fatty
acids in turn form water, carbon dioxide and methane. The alcohols
can undergo oxidation to aldehydes and ketones.

Ammonia may react with alcohols and organic acids to produce
amines and amides. Hydrogen sulphide can react with alcohols and
organic acids resulting in mercaptans and thioacids. Further reactions can produce disulphides. Bacterial metabolism of the inorganic portion of manure can also produce ammonia, hydrogen sulphide
and carbon dioxide.[18]

An entire chemical soup, from pig shit. Their health effects, like
those of industrial chicken farm waste, can be severe, as a report by
the North Carolina Department of Health and Human Services
explains:

Exposure to environmental odors [from hog farms] results in physio-
logical stresses that may result in a variety of symptoms including
headache, nausea, loss of appetite, and emotional disturbance.
Odors may exacerbate stress-related illnesses. The symptoms may
result from odor annoyance, stress associated with odor exposure,
and conditioned responses to odors. The literature also reports that
exposure to odors may exacerbate asthma symptoms[19]

Those who work inside such facilities can suffer still worse effects:

Many studies have reported the health effects experienced by people working in hog confinement buildings. Donham et al. (1989) reported the following acute symptoms and prevalence rates in a study of hog confinement workers:

- *Cough (67 percent)*
- *Phlegm production (56 percent)*
- *Scratchy throat (54 percent)*
- *Runny nose (45 percent)*
- *Burning and watery eyes (39 percent)*
- *Headaches (37 percent)*
- *Chest tightness (36 percent)*
- *Shortness of breath (30 percent)*
- *Wheezing (27 percent)*
- *Muscle aches and pain (25 percent)*

Schwartz et al. (1990) reported that chronic effects are manifested as bronchitis, where airway obstruction was present affecting up to 25 percent of hog house workers. Long term lung damage may occur as pulmonary function tests indicate air trapping in lungs and a persistent leukocytosis [increased white blood cell count].[20]

Both chicken and hog farm wastes have been associated with alarming recent outbreaks of Pfiesteria piscicida, a one-celled animal sometimes referred to as "the cell from hell." As Rick Dove writes, these outbreaks have often been marked by mass fish kills in rivers or creeks downstream from industrial farm operations:

In the 1995 fish kill [on the Neuse River, near New Bern, North Carolina] for over 100 days fish were once again dying in large numbers. Nearly all of them were covered with open bleeding lesions. In just 10 of those 100 days, volunteers working with the Neuse River Foundation documented 10 million dead fish. At that time, many cit-

izens who were exposed to this fish kill complained about a number of neurological and respiratory problems. North Carolina health authorities documented these problems and wrongly dismissed them. Later, researchers working similar fish kills on Maryland's Pokomoke River would link these same symptoms to the cause of the fish kills, Pfiesteria piscicida.

This creature produces an extremely powerful neurotoxin that paralyzes the fish, sloughs their skin, and eats their blood cells. It is capable of doing the same thing to humans. This neurotoxin is volatilized to the air and is known to cause serious health problems, including memory loss, in humans who breathe it. Its proliferation has been directly linked to nutrient pollution from concentrated animal feeding operations. [21]

"The fish kills continue today," Dove continues. "Depending upon weather conditions, some years are worse than others. Many smaller kills are not even counted. Fishers [the politically correct term for fishermen and women] continue to report neurological and respiratory symptoms, and a dark cloud still hangs over the state's environmental reputation and economy."[22]

Far more horrifying than even the *Pfiesteria* problem was the spring 2000 outbreak of *E. Coli* bacteria in the drinking water of the Canadian town of Walkerton, Ontario, which killed at least six people and made an additional 2,000 ill. A report submitted to the Ontario legislature by Environmental Commissioner Gord Miller pointed the finger at the region's industrial farms. *The Canadian Press* wrote:

Coinciding with a public inquiry into the water-borne E. Coli outbreak that infected 2,000 people and contributed to six deaths at Walkerton, Miller's report criticized lack of regulation of large farms that produce vast amounts of liquid manure.

He said the situation has made Ontario a "haven" in North America for companies wanting to avoid rigorous controls. "In many other jurisdictions, including Quebec and the U.S., there are laws and regulations governing the management of manure. But in Ontario there is virtually no control," he said.

The report proposed tightening industry standards and subjecting large farming operations to an approval process.[23]

Concludes Dove:

*[Chicken and hog] factories do not produce meat more efficiently than traditional family farmers. The industry's willingness to treat the animals with unspeakable cruelty and to dump thousands of tons of toxic pollutants into our nation's waterways, **and their ability to get away with it,** however, has given it a dramatic market advantage over the traditional family farm. Indeed, the industry's business plan is based upon its ability to use its political clout to paralyze the regulatory agencies, thereby escaping the true costs of producing its product."*[24]

WILDLIFE HABITAT DESTRUCTION

The threat posed to wildlife by factory farming is not limited to insects and the underwater world, nor traceable only to pollution. Of greater concern is the loss of habitat as more land is brought under cultivation, and land already cultivated is consolidated in ever larger monocrop operations.

In contrast to natural forest, which seen from the air presents a relatively solid expanse of green, regions given over to farming are characterized by "mosaic landscapes," an irregular checkerboard pattern of varicolored fields and woodlots connected by zig-zagging fencelines. For insects, birds, and mammals on the ground, the helter-skelter squares constitute a complex jumble of micro-environments, with fencerows linking them.[25] Under traditional mixed farming regimes, the jumble was complicated enough that it often benefited wildlife, providing shelter and/or forage for different species at different seasons. In North America, for example, the expansion of agriculture in northern New England and parts of Canada caused a decline in woodland moose populations, but resulted in a veritable population explosion among whitetail deer—which eat corn and find the open, mixedfarm environment ideal. The same has been true of various insect, rodent and bird species.

The landscapes created since the Second World War by industrial farming represent a marked, in some cases radical, environmental simplification, with negative results for most wildlife species. Britain provides an example of the effects on birds, examined in *Farming and Birds*:

> *Mixed farming systems, combining tillage crops and stock, frequently emerge in our analyses as most favorable to birds. Yet farming has changed dramatically over the last 40 to 45 years, from an industry whose methods were relatively favorable to wildlife to a specialized, highly technical business heavily biased against the maintenance of the diversity of nature.*
>
> *Modern farming is the antithesis of the image projected by the old adage about sowing corn—one for the rook; one for the crow; one to rot and one to grow. Nowadays the farmer sows [only] one to grow; competition within the industry allows little margin for natural wastage. (Wright, 1980).*[26]

Farming affects bird populations in two ways:

> *The first is by engulfing and totally destroying particular habitats within farmland, such as woodlands and lowland heaths. The second is by modifying the nature of the surviving habitats, thereby altering the niches they offer in ways that make them less (or perhaps occasionally more) attractive to birds.... Since 1939 the conversion of permanent pasture to arable production has constituted a major loss of habitat.*[27]

Engulfing usually occurs when the land is first cultivated. In Britain, this may have been hundreds of years ago. Industrial farming's much more recent effects are felt in the form of "modifying the nature of surviving habitats," though in some cases "modifying" should be replaced by "eliminating." The changes are most often due directly to mechanization and farm expansion. Write O'Connor and Shrubb:

> For modern arable farms to be viable with low labor inputs, farm structure must allow the most efficient use of large machines. The impact of such machinery on farmland practice [is] magnified by the growth in the average size of farm holdings in Britain. Between 1875 and 1979 farm sizes have increased by a factor of 2.0-2.5.[28]

As for the effects on birds:

> The increasing size and speed of machines, which tend much more to divorce the operator from his immediate surroundings, may well increase nest and brood losses in ground-nesting species. Second, and probably more important, are the effects of mechanization in narrowing the complexity and timing of crop rotations.[29]

On a traditional farm, where as many as six to ten field crops may have been combined with garden vegetables, pasture, and livestock, crop rotations were frequent, often complex. As O'Connor and Shrubb note, "the classic four-course Norfolk rotation [red clover (*Trifolium pratense*), winter wheat, sugarbeet, spring barley] not only provides cleaning crops (roots) and compost or manuring crops (clover) but also provided for the most economic use of labor by spreading the farm's work fairly evenly throughout the year."[30] Under industrial farming, however:

> The bulk of the crop is now planted in autumn, despite the workload this imposes. Formerly the work used to be divided fairly evenly

between spring and autumn. The resulting change in the timing of cultivation has also been promoted by the decline in root crops in many areas and the increasing concentration on cereals.

Changes in the periods of tilling the soil have a significant impact on the availability of food supplies to birds, particularly in the breeding season. Spring cultivations are a major source of invertebrate food for many birds when the start of breeding may be limited by the availability of food. Autumn-sown fields are unsuitable feeding areas in spring since the vegetation is by then too tall and dense. Ground-nesting birds such as Lapwing may also prefer spring cultivations for nesting.[31]

The removal of fencerows to consolidate farm fields (as our late, unlamented neighbor Bolduc did on his farm) also destroys bird habitat, and cuts off the sheltered connections between fields along which rodents and other small mammals—which make up the diet of most hawks and owls—travel. Drainage of wetlands, and filling in of drainage ditches between fields, reduces habitat for ducks and other waterbirds. An example of how great the impact of human activity on birds can be is the greater prairie chicken. Once common on the western plains, it has been extirpated from Canada since the turn of the century, largely due to habitat loss caused by human settlement.[32]

What is true for birds is true for other species, in locations as far apart as Western Canada and East Africa.

On the Canadian prairie, writes David Wylynko:

Loss of habitat [due to agriculture] is cited as the primary reason for the decline in most prairie species. During the 1900s, four mammals—the grizzly bear, wolf, black-footed ferret, and swift fox—have been extirpated from the Canadian prairies. Three more—the bobcat, long-tailed weasel, and badger—are at considerable risk.

Both the number and abundance of bird species have decreased dramatically in the transition from native habitat to cultivated land

*and tame pastures as well. Among birds that breed in or migrate
through the prairies, more were placed on Canada's list of species at
risk in 1994 than in any other ecozone in the country.* [33]

Modern agriculture has been identified as an indirect threat to the
future of the burrowing owl and prairie falcon, because of farmers'
dislike for these species' main food source: Richardson's ground squir-
rel. Ground squirrels, or "gophers," are considered vermin and exter-
minated by farmers because their varied diets include wheat, barley
and a number of popular garden vegetables, while their burrows pose
dangers for cattle that step in them. Though not themselves endangered, the gophers' elimination destroys a forage source for much
rarer predators that eat them.[34]

Encroachment of human settlement and the "development" of
former wilderness or semi-wild areas for ski resorts and golf courses
has led to other, unforseen environmental problems. These range
from aggressive elk terrorizing tourists in Colorado, to cougar attacks
on joggers in British Colombia, to the spread of Lyme disease by the
burgeoning deer populations of New England, the U.S. Midwest, and
the Pacific coast.

A multisystem disorder caused by the spirochete *Borrelia burgdor-
feri*, Lyme disease is carried by ticks that feed on deer, then hide in
grass and subsequently bite humans. In areas where the natural predators of deer have been wiped out, and where hunting by humans is
banned, deer herds often increase to the point of becoming serious
pests, even in heavily populated big city suburbs. Proposals to cull the
overpopulated herds by extending hunting seasons or ranges often
meet with outrage from urban people conditioned to thinking of all
deer as so many Walt Disney Bambis.

The problem is fairly recent in North America, but it has a proto-
type—and possibly prototypical solution—in another part of the
world where farming impinges on the wild. In Africa, farming is likely the most serious potential danger—more serious in the long term

than that of today's ivory poachers—to *Loxodanta africana*, the African elephant. While international attention is focused on high-profile anti-poaching "wars" and public burning of impounded ivory, the threat to elephant survival posed by habitat loss due to the expansion of agriculture and grazing goes comparatively unreported. Yet the threat is clear, as one recent statement of the economic aspects of the problem reveals:

> *The competing for land argument embraces Hardin's notion of the survival of the fittest, where the demands of two sympathetic species are sufficiently similar that competition between them leads to the extinction of one. In this case, the species are humans and elephants, who compete for essential resources of food and habitat. Humans' plant food demands are similar to those of the elephant, and they also indirectly compete for the use of the same resources for their domestic stock. This competition is likely to be significant, given that the African population doubling-time is now merely 18 years, and has brought about rapid forest conversion for agricultural and pastoral activity.*[35]

This basic economic cause for competition between the two species is accentuated by the tendency of Africans not to consider any long-term benefits that might accrue from the conservation of wildlife. As the study authors note:

> *Rural people in Africa are likely to prefer income in the present as opposed to the future. In Africa, average life expectancy at birth is 51 years and infant mortality is 10 percent. These figures are even more grim in rural areas where health care, nutrition, education, and clean water are scarce. Thus risk of death is a crucial factor. Uncertainty of the future is also compounded by threats of drought and other natural disasters, political instability and warfare, economic disruptions and policy changes and natural resource degrada-*

tion. Finally under conditions of extreme poverty, a household's major concern is securing sufficient means for survival today. Observers suggest these conditions leading to a high rate of time preference are reflected in the lack of sustainable farming practices (practices which conserve soil quality, irrigation water, and rainfall) in sub-Saharan Africa. This implies that individuals are not willing to bear the risk and uncertainty from changing farming patterns in order to benefit from future gains.[36]

Nor are they willing to accept the present checks of being kept out of game areas when they want to expand their land holdings, or of elephants destroying crops on already-cultivated fields. As a villager in Zimbabwe once observed: "Wildlife is nothing but a nuisance. Elephants destroy our crops every night. They (the government) can kill everything bigger than a hare as far as we are concerned."[37]

A solution to the conflict exists—one that protects both elephant herds and the financial interests of African farmers and rural villagers—and has been successfully field-tested in Zimbabwe. But whether the Communal Areas Management Program for Indigenous Resources (Campfire) will be widely adopted by African governments is in doubt.[38]

Launched in 1986, the program is based on two principles: 1) the concept that wildlife is an economic resource with real value to the rural economy and, 2) the idea that local resources should be controlled locally. In the past, most African governments followed the game conservation pattern of colonial times, that of setting aside reserves for wildlife hunting and tourism, and keeping local rural people away—sometimes even evicting them from parklands. As African populations grew and pressure to put more land—even marginal land better suited for wildlife—under cultivation, rural people began to resent being barred from reserves. Squatting—illegal occupation of land—and poaching increased in some countries to the point of "virtual warfare" between rangers and local villagers.[39]

Realizing a crisis had been reached, Zimbabwe tried a new approach. Management of wildlife in communal land areas was given over to local district councils, which adopted a quota system. An optimum number of wildlife that could be supported in a given habitat was set, and surplus animals, as well as those which caused damage to nearby crops, were culled. The meat, hides, and ivory were sold and earnings were used partly to pay for wildlife management (including anti-poaching patrols) and partly returned to local villages for use in such development projects as school and hospital construction, and for cash distribution to local households. As the deputy director of the Zimbabwe Department of National Parks and Wildlife Management notes:

> Government has issued a set of guidelines to councils suggesting that a minimum 50 percent of total income should be returned to the wards and villages where it was generated. A maximum 35 percent should be reinvested in wildlife management costs for the following year and a maximum 15 percent should be retained by the council for administrative purposes.
>
> From recent studies of the wildlife industry in Zimbabwe, it seems the net financial returns from land under wildlife significantly exceeded those possible from cattle (US$1.11 per hectare versus $0.60 on commercial farms in Natural Region IV of Zimbabwe), and the potential for improvement in wildlife returns is far greater than that for cattle (up to $5 per ha for hunting and up to $25 per ha for ecotourism). For the past 20 years the amount of land allocated by landholders to wildlife has been increasing and, including state-protected areas, almost one-third of Zimbabwe is now under wildlife. The trend is likely to continue as marketing of wildlife improves.[40]

As a result of this innovative program, the elephant herd in Zimbabwe actually increased during years when that in nearby Kenya, with a protectionist philosophy, plummeted dramatically. Despite its success and popularity with farmers, however, the Campfire program has been attacked in Zimbabwe and abroad. Internally, R. B. Martin writes:

> *Campfire is seen as a threat: Rural people have organized themselves into suitable groups and established committees to undertake wildlife management. These institutions are becoming vociferous in the political arena. If their representatives, both on local committees and at the level of members of Parliament, don't "produce the goods," others are elected in their place. Unexpectedly, Campfire has become the engine for a powerful democratic movement among remote communal land people. As such it is a threat to politicians and bureaucrats who do not really wish to see self-sufficient rural communities."*[41]

Outside Africa, the program—because it permits hunting and the sale of ivory—is viewed by some environmental and animal rights groups as heresy:

> *Certain extreme right wing international "green" organizations perceive the program as a threat to their conservation ideologies. If it can be demonstrated that wildlife populations increase through sustainable use programs, this weakens international stands against any form of wildlife exploitation by humans.*[42]

An international effort has been launched in the northern industrial region to discredit the Campfire approach, and many donor countries are reluctant to offend constituents by allocating funds for similar efforts in other African nations. The eventual issue of this battle of conservation philosophies could well decide the fate of the African elephant, and provide a model—good or bad—for North America as well.

SOIL DEGRADATION

Industrial farming, especially when carried out carelessly and without due regard to local environmental characteristics, can do great harm to soils, promoting topsoil erosion, soil compaction, salinization, and loss of fertility. Agriculture Canada researchers warn the damage is often not immediately apparent:

> *Problems develop only over long periods. The short-term, negative effects of various farm management practices on the soil are often hard to identify. In the long-term, cumulative effects reduce soil productivity. When nutrient-rich topsoil is lost, the problem can be overcome in the short term by increasing fertilizer input. However, this adds to production costs. Other forms of degradation are less easily remedied. For example, using more fertilizer does not increase yields if plant growth is limited by compacted soil layers hindering root development.*
>
> *Besides the negative effect experienced by the farmer, soil pollution and soil degradation also have long-term implications for society. Not only is the land's potential for food production reduced, but other large-scale environmental impacts, such as the spread of desert lands and flooding, could occur.*[43]

A number of practices common in industrial farming—excessive tillage, summer fallowing, and razing windbreaks to permit expansion and consolidation of fields—increase wind and water erosion, causing various noxious effects:

> *Besides contributing to pollution, soil eroding onto other properties creates a nuisance. It can also clog drainage and irrigation channels. Wind and water erosion removes the fine topsoil and associated nutrients from farmlands. If not controlled, it decreases the productivity of land as the subsoil content of the plowed layer increases. As subsoil is incorporated into the cultivated layer, soil fertility and water-holding capacity are reduced. Root growth and development are also limited, resulting in variable crop growth. Soil transported*

by erosion and redistributed within the field can increase localized ponding, smother young seedlings, and cause a crust to form on the surface after drying. In severe cases erosion produces such extensive gullies that workable land is lost completely.[44]

Eroded soil that enters waterways becomes sediment that eventually fills in shipping channels, clogs hydroelectric dams, and causes fish die-offs. In the Great Lakes basin, dredging of harbors to clear them of silt and keep shipping lanes open costs in excess of $100 million annually.[45]

Soil compaction, caused by poor tillage practices—including the use of big machines whose weight compresses the soil (a penchant of farmer Bolduc)—brings its own problems:

A few soils are naturally dense and limit plant growth. Man can create similar, undesirable conditions by working soil when it is wet and using machinery with excessive weight or at excessive speeds. Repeated cultivation can also increase oxidation rates and microbial decomposition. These phenomena reduce the organic matter content, making it easier for tillage to pulverize soil aggregates and destroy soil structure. Loss of structure makes the soil more susceptible to erosion and further compaction.[46]

Under an industrial agricultural regime, time is money, and speed in performing tillage tasks is of the essence. Large machines are operated at maximum speed by hired hands, who race through their work, with little attention to the long-term effect on soils. Debts must be paid down every month, regardless of the biological needs of soil organisms. The eventual effect on the source of life itself is not reflected in the mortgage interest tables.

Irrigation, an indispensable tool of modern agriculture, can also threaten soils through salinization. When too much water is used to irrigate fields, the water table can be raised, causing soluble salts in the earth to rise with it. When the water evaporates, the salts are left in surface soil layers. Warns Agriculture Canada: "Contamination of the soil with salts reduces germination and growth of many crops. Increased sodium levels cause deterioration of soil structure by inducing the formation of dense soil layers that restrict water movement."[47] Leached salts in groundwater can bring problems far from the farmland where they originated, increasing the salinity of rivers that may be the only source of drinking water for downstream municipalities.

Some of the most glaring examples of the problems posed by salinization have been observed in that pillar of the industrial farming system, California. There, as already briefly mentioned in the previous chapter, irrigation is an absolute agricultural necessity, especially in arid or semi-arid parts of the state. And irrigation by its very nature tends to create problems with salt pollution that have impacts far beyond the irrigated land itself.

As David Carle explains, "Crops separate irrigation water from most of the salts it carries, leaving the minerals concentrated in the soil."[48] To keep that salt from building up to the point where crops can no longer survive, farmers must flush it away. To do this, they apply more irrigation water than the plants themselves actually need to grow, which excess water "flushes salts down into the groundwater below."[49]

At this point, farmers can get into trouble if the earth beneath their fields happens to include an impermeable layer, blocking the salts from flushing away. All too often, due to modern industrial tilling systems, there is just such a "hardpan" layer at a depth immediately below the level where a plowblade would reach. Continual passages over the fields with heavy farm equipment (like Bolduc's huge tractor) pack down the soil, and if the plowblade can't reach down far enough to break it up, hardpan results. The salts have nowhere to flush.

Farmers try to avoid this problem by "tiling" their fields, namely, by burying perforated pipes in the ground at spaced intervals beneath the ground. These pipes, or "tiles," carry the salts away. However, even if over-tillage doesn't create a hardpan layer, or tiling allows the fields to flush more easily—thereby permitting the farmers themselves to avoid salt pollution—the salts flushed out of the fields can wreak havoc elsewhere. Basically, industrial farmers may only be passing the buck to people downstream.

Carle explains:

> The San Luis Drain [serving the San Joaquin Valley] was planned to carry water northward through the valley all the way to Suisun Bay. Eighty-five miles of the drain were built, beginning in the south. By 1973, the drain reached as far as Kesterton National Wildlife Refuge, near Gustine. Concerns about the impacts of sending the polluted water into the Delta halted construction, forcing the water to remain at Kesterton. In 1983, after a decade of drainage water ponding there, thousands of waterfowl and shorebirds began to be born with deformities and to die. Bird embryos had protruding brains, missing eyes, twisted bills, and grossly deformed legs and wings.[50]

"Even without the contributions of the San Luis Drain," Carle continues, "San Joaquin River flows that approach the Delta today consist of little but agricultural drainage water.... Salts, inevitably, will continue to concentrate. After several decades, the farmers will have to deal with permanent storage of very concentrated slurries."[51]

An example of how serious a problem "concentrated" salt pollution can create can be seen in California's famed (or perhaps infamous) Salton Sea, which collects drainage water from farms in the Imperial Valley, where "38,000 miles of subsurface drainage pipes gather water from 500,000 irrigated acres."[52] Carle continues:

> The Salton Sea is the "evaporation pond" for that water, taking in more than four million tons of dissolved salt and tens of thousands of tons of fertilizers each year. It is now 25 percent saltier than the ocean.
>
> Salinity levels are approaching the limits for any fish. If no

*freshwater entered to replace evaporation, the sea could lose its fish
in two to eleven years..*

   *Exacerbating efforts to address the Salton Sea's progressive salina-
tion, coastal cities have plans to purchase water from Imperial Valley
farmers to service more population growth and development. Although
the farm drainage water gradually increases salinity in the Salton Sea,
losing Colorado River water to those cities would cause a more rapid
decline in water level and further concentration of the salts.*[53]

The mere act of "watering the plants," when performed on an
industrial scale, can bring amazing difficulties and cause wildly dispro-
portionate damage.

WASTE OF FRESHWATER RESOURCES

The globe has enough fresh water to supply the needs of everyone,
but much of that water isn't available when and where needed. Sandra
Postel, director of the Global Water Policy Project, notes:

   *At least 20 percent of the renewable water supply—generated each
   year by the solar-powered hydrological cycle—is too remote from population centers to be of use. A large portion runs off in floods, and
   cannot be supplied reliably to farms, industries and households. After
   accounting for this unequal distribution of water in time and space,
   the picture of plenty is revealed to be an illusion."* [54]

Clashes over water rights are becoming increasingly common,
both within countries and internationally, and when fingers are pointed at those deemed responsible for shortages, farmers are inevitably—
and justifiably—singled out: Agriculture accounts for two-thirds of all
water drawn from rivers, lakes, and aquifers. If the massive amounts
of fresh water used in farming are not used wisely or actually degrade
the water supply (i.e. through salinization), political and even military
conflict can result.

And much of the water used for agriculture does, in fact, result in waste of the water itself and degradation of other resources. Every year, up to 10 percent of irrigated lands lose their productivity due to salinization caused by poor water management.[55] Damage on a much greater scale has been caused by the construction of large dams and diversion schemes on the world's major rivers, inspired in large part by agricultural demands. As Postel points out:

> Globally, water demand has more than tripled since mid-century, and it has been met by building ever more and larger water supply projects, especially dams and river diversions. Around the world, the number of large dams (more than 15 meters high) climbed from just over 5,000 in 1950 to roughly 38,000 today. More than 85 percent of large dams have been built during the last 35 years. Many rivers now resemble elaborate pumping works, with the timing and amount of flow completely controlled by planners and engineers.
>
> Unfortunately, this massive manipulation of the hydrological system is wreaking havoc on the aquatic environment and its biological diversity. River deltas are deteriorating, species are being pushed toward extinction, inland lakes are shrinking, and wetlands are disappearing.
>
> In the Aral Sea basin, site of one of the world's worst environmental tragedies, what was once the planet's fourth largest lake has lost half of its area and three-fourths of its volume because of excessive river diversions to grow cotton in the desert. Some 20 of 24 fish species have disappeared, and the fish catch—which totalled 44,000 tons and supported some 60,000 jobs in the 1950s—has dropped to zero.[56]

An even greater disaster has been visited upon Egypt, where dam construction was ironically intended to boost farm production, as well as provide electrical power:

> The Aswan High Dam on the upper Nile, completed in 1963 and fully

effective by the 1970s, was intended to regulate the river's flow and provide hydroelectric power. The traditional fertility of the Nile Valley, however, was dependent on more than 100 million tons of silt deposited annually as the river flooded. The silt is now silting up the artificial Lake Nasser, forcing farmers downstream to rely on fertilizers and robbing local brickmakers of raw materials; about 35 percent of Egypt's cultivated land is suffering from salinization. Deprived of nutrients, the fish stocks of the eastern Mediterranean are declining, while the Nile delta is being eroded steadily. Schistosomiasis, a disease common in newly irrigated areas, has spread explosively, causing debilitation and death. In 1990 between five and six million people were affected.[57]

At international level, decisions by governments as to how to use the water from rivers that feed not only their own but also neighboring countries have resulted in serious inequities. An example is the Ganges River, which flows through India and Bangladesh. As Postel notes:

Bangladesh suffers from India's unilateral diversion of the Ganges at the Farakka barrage, and the failure, since 1988, to reach agreement on how to share the Ganges during the dry season. In 1993, the dry season flow into Bangladesh was the lowest ever recorded. As riverbeds dried up and crops withered, the northwestern region suffered greatly. The Ganges Kobadak Project, one of this poor nation's larger agricultural schemes, reportedly suffered $25 million in losses.[58]

Other river systems where international conflict over freshwater use rights has reached crisis level include the Danube River basin, shared by 17 squabbling countries;[59] the Nile River basin, whose waters are disputed by Egypt and Sudan; and the Tigris-Euphrates River basin, over which Turkey, Syria, and Iraq threaten to come to blows.[60] In each region, agricultural use is a leading bone of contention.

In North America, similar conflicts are starting to build. There have been numerous legal and political battles, for example, over the diversion of water from the Colorado River, whose watershed area includes no fewer than seven U.S. states, as well as Mexico, and which no longer empties its waters into the ocean via the Gulf of California because every drop in it is used up by farms or cities along its route.

Agriculture has for decades been the heaviest user of the Colorado's water. Without it, California's agriculturally crucial Imperial Valley could not continue producing its bumper crops. But the time when agriculture took easy precedence over other water uses in California may be drawing to a close, as the state's burgeoning urban conglomerations clamor for a greater share. As Carle writes: "the forced transfer of agriculture water to serve urban interests seemed to be a sign of California's future."[61] Plans have been suggested to decrease the use of irrigation water from the Colorado by allowing some farm fields to lie fallow (temporarily unused for cropping). But this may only complicate salinization problems, particularly those of the Salton Sea. If less water is diverted through the system, the irrigation flows to the sea that flush excess salt would be cut back, and salt concentration increased.

A potentially even greater problem, with international implications, is being created by the exhaustion of underground freshwater aquifers in the American West. This has already prompted legislators to discuss the idea of expropriating the contents of the Great Lakes as a source of irrigation water for U.S. industrial megafarms. The fact that the international boundary between the U.S. and Canada runs directly through those same Great Lakes, and that Canada would have to be consulted and placated before wholesale water diversion could begin, is seen as an annoying irritant by some U.S. politicians. Canadians, in turn, view U.S. ambitions for the lakes with alarm, realizing that the ecological and economic effects of draining them could create a major disaster. The Great Lakes commercial and sport fishing industries—a lucrative source of employment for the people of Michigan and Wisconsin, as well as Canadians—could be destroyed, with the resulting loss of livelihood and recreation for thousands of people.

The greater the thirst of agribusiness's industrial megafarms, the more problems are created.

LOSS OF BIODIVERSITY

Under traditional agricultural systems, farmers keep some grain from every harvest as seed for the next planting season. In each ecological region, sometimes after centuries of trial and error, they have selected plant varieties suited to their all-round needs: those best adapted to local weather and soil conditions, most resistant to local pests and plant diseases, and most productive of the wide variety of products needed on a typical mixed family farm. They choose varieties of rice, wheat, barley, millet, or sorghum that produce not only enough grain for human consumption, but also straw and leaves for livestock feed, as well as for use as mulch or "green manure" on their fields. The number of available varieties is surprising. For example, FAO researchers in Ethiopia stopped to examine a local farmer's two-acre plot—and identified 11 varieties of indigenous wheat, which the farmer had planted for a range of specific uses.[62]

Of the world's estimated 500,000 plant species (roughly half of which have been scientifically classified),[63] farmers have domesticated hundreds which, when multiplied by the number of varieties developed within each species, provide a panoply of planting options capable of adapting to almost every environmental eventuality. As this author noted in a published interview with Murray Bookchin:

> Paraphrased, Bookchin's watchword might be expressed: in diversity there is strength. The dynamic tension maintained between a variety of organisms in an ecosystem gives the system its resilience. The more forms it takes, the less likely life is to be wiped out by fire, flood or Ice Age climate changes.
>
> The traditional mixed farm, producing a variety of crops, pos-

sessed the same strength. If hog prices were down, a good corn crop might take up the slack. An abundant maple sugar run could offset the loss of a Jersey calf.[64]

In modern, industrial farming, however, the goal of plant breeding is turned on its head: Diversity as a trait is considered undesirable and deliberately "selected out." A single aim—most frequently maximum grain production for immediate export—takes precedence over all others, and only crop varieties which contribute directly to it are favored. Plants with long stems, which might be useful on a mixed farm as straw, are discarded in favor of those with shorter stems and larger grain heads. Varieties which grow to the exact height required for fast, mechanical harvesting are favored over shorter or taller strains that might be resistant to local insects. (It is assumed that pesticides will protect the more vulnerable plants.) Farmers cease to save the seed of older, local plant varieties, which may as a result be lost forever. To assure processing factories of "uniformity of product," only the new varieties are planted—row on row, acre after acre—in huge mono-crop operations:

> With everything invested in hundreds upon hundreds of acres of a single uniform crop, a tiny shift in market price would mean disaster if the system were not supported by government intervention. Without the artificial protection of thousands of dollars worth of harsh chemical pesticides, the fragile monocultural crop could also be wiped out overnight by an invasion of insects attracted by the unnaturally rich feast spread before them.[65]

Such a major pest invasion—in this case a fungal infection rather than insects—has, in fact, already occurred, as Baeza-Lopez reports:

> *The intrinsic weakness caused by the genetic uniformity of modern varieties was underlined in 1970, when the fungus*

Helminthosporium maydis attacked maize in the United States. Production dropped by 15 percent nationwide and 50 percent in the affected areas, causing losses of millions of dollars.

The commission charged with investigating the maize disaster concluded the cause was genetic uniformity. Nearly all hybrid varieties in the country had been obtained from a single source of parental sterility called Citoplasm texas, *which was susceptible to this new form of fungus.*[66]

The lesson seems to have been lost on the industrial farm sector. Instead of a wide selection of local crops, the system continues to prefer the types of plants bred by the scientists who created the Green Revolution: high-yielding, "improved" grains. Most often these are hybrids which, because they are the offspring of genetically dissimilar parents, cannot "breed true" in subsequent seasons. Farmers cannot save the seed from such crops to plant again, but must purchase new stock each year from seed companies. Indian scientist Vandana Shiva analyzes the situation:

The miracle of the new seeds has most often been communicated through the term "high-yielding varieties (HYVs)." The HYV category is central to the Green Revolution paradigm. However, unlike what the term suggests, there is no neutral or objective measure of "yield" on the basis of which the cropping systems based on miracle seeds can be established to be higher yielding than the cropping systems they replace.

The Green Revolution category of HYV is essentially a reductionist category which decontextualizes properties of both the native and the new varieties. Through the process of decontextualization, costs and impacts are externalized and systemic comparison with alternatives is precluded.

Since the Green Revolution strategy is aimed at increasing the output of a single component on a farm, at the cost of decreasing other components and increasing external inputs [e.g., inorganic fer-tilizers], such a partial comparison is by definition biased to make the new varieties "high-yielding," when at the systems level they may not be.[67]

Thus a variety of wheat that produces abundant, large seed heads with a high protein content—but which has short, weak stems of lit-tle use as straw, is prone to lodging (wind damage), and is perhaps vul-nerable also to local insect, bacterial, or fungal pests—is labelled "improved." The fact that the variety may require large doses of cost-ly inorganic nitrogen fertilizers and frequent applications of equally costly pesticides to survive—and that meeting these costs draws farm-ers into a debt spiral—is not considered relevant. Nor is sufficient attention paid to the fact that, in allowing local crop varieties to be dis-carded and die out, plant breeders are depriving themselves of the very sources of fresh germplasm their programs might need to meet as yet unforseen future needs. Shiva continues:

The indigenous cropping systems are based only on internal organic inputs. Seeds come from the farm, soil fertility comes from the farm and pest control is built into the crop mixtures. In the Green Revolution package, yields are intimately tied to purchased inputs of seeds, chemical fertilizers, pesticides, petroleum and to intensive and accurate irrigation. High yields are not intrinsic to the seeds, but are a function of the availability of required inputs, which in turn [may] have ecologically destructive impacts.

In the breeding strategy for the Green Revolution, multiple uses of plant biomass seem to have been consciously sacrificed for a single use, with non-sustainable consumption of fertilizer and water. The increase in marketable grain has been achieved at the cost of decreased biomass for animals and soils and the decrease of ecosys-tem productivity due to overuse of resources.[68]

Critics of various Green Revolution projects stress the fact that the concentration on hybrid seed, which must be re-purchased each year, leads to farmer dependency. One such project, conducted by Japan's Sasakawa Foundation in Ethiopia, has been described as:

> *an artificial laboratory. SG 2000 selects the crops and technical packages without farmer consultation, procures the inputs and delivers them to the participating farmers via the village extension agent who also handles credit recovery. The packages are high-input/high-output/high-risk and create farmer dependency on imported hybrid seed (Pioneer brand) and fertilizer.*[69]

Farmer dependency becomes particularly ominous when it is realized that the international seed trade is becoming increasingly concentrated in the hands of a small number of companies—often the same chemical manufacturing firms that produce the fertilizer and pesticides the new, "improved" seeds need to survive. Twenty years ago most seed sold to farmers came from long-established, local seed houses. Today, more than 30 percent of the seed sold in the northern industrial countries is controlled by 20 large firms, many of them pharmaceutical or chemical producers. The top six in terms of sales are Pioneer Hi-Bred, Sandoz, Limagrain Inc. Nickerson, Upjohn, ICI, and Cargill.[70]

The possibility that, if present trends continue, the majority of the world's farmers could become totally dependent for seed, fertilizer, pesticide, herbicide—even for the very structure of their increasingly simplified and ecologically vulnerable farming systems—on a small group of transnational corporations whose ultimate *raison d'être* is not food production, but financial profit, and that the number of crop varieties available through these firms represents a drastically impoverished ecological potential, ought to give pause to planners everywhere.

It might well give them cold sweats, if the risks involved in the new processes of plant and animal bioengineering are factored into the equation. As the Environmental Defense Fund's Rebecca Goldburg explains:

> With traditional selective breeding, humans can cross one crop variety, say of potatoes, with another potato variety, and in some cases with related wild potatoes. But traditional breeders cannot add viral, insect or animal genes to potatoes—all of which have been added to plants via use of "recombinant DNA" and related genetic engineering techniques. Scientists can now, at least in theory, take virtually any genetically encoded trait from one organism and add it to another, no matter how unrelated
>
> For those concerned about the environment, the prominent issue has been the possibility that scientists will inadvertently create "transgenic" organisms that wreak ecological or other types of havoc, particularly if [they] are intended for deliberate release into the environment.[71]

EXOTIC SPECIES

Examples of the kind of mischief that can result when exotic organisms are released into environments where they have no natural ecological niche, and often no "control" in the form of local predators or competitor species, abound in history, particularly the story of European exploration and colonization. Alfred W. Crosby, in *Ecological Imperialism: The Biological Expansion of Europe, 900 through 1900*, documents many disasters in the 1,000-year span of expansion of European civilization. For example, he describes the colonization of Porto Santo and Madeira by the Portuguese:

> Madeira and Porto Santo were virgin in the purest sense of the word. They were uninhabited and bore no mark of human occupation, Paleolithic or Neolithic or post-Neolithic. The newcomers set to work to rationalize landscape, flora, and fauna previously unaffected by

*anything but the blind forces of nature. Bartholomeu Perestrello, cap-
tain donatory of Porto Santo (and, incidentally, future father-in-law
of Columbus), set loose on his island where the likes of such had
never lived before a female rabbit and her offspring; she had given
birth on the voyage from Europe. The rabbits reproduced at a villain-
ous rate and "overspread the land, so that our men could sow nothing that was not destroyed by them."*

*The settlers took up arms against these rivals and killed great
numbers, but in the absence of local predators and disease organisms
adapted to these quadrupeds, the death rate continued to lag behind
the birth rate. The humans were obliged to leave and go to Madeira,
defeated in their attempt at colonization not by primeval nature but
by their own ecological ignorance. Europeans would make such mis-
takes over and over, setting off population explosions of burros in
Fuerteventura in the Canaries, rats in Virginia in North America,
and rabbits [again] in Australia.*[72]

More recent examples include introduction of the European starling (*Sturnus vulgaris*) to New York City in the late 19th century, and of
the infamous cane toad (*Bufo marinus*) to Queensland, Australia in the
20th century.[73] The starling, introduced for aesthetic reasons, reproduced at a prodigious rate, displacing and nearly wiping out the native
North American bluebird and creating a nuisance with its huge noisy
flocks. The cane toad, which is poisonous, was introduced to help control sugarcane beetle, but instead destroyed local Australian wildlife
species and reproduced so rapidly as to become a virtual plague.

Plant species have exhibited similar behavior, for example various
crops transported by the Spanish to California and Peru:

*[The Spanish] did write about respectable plants that went wild and
defied attempts to keep them out of cultivated fields, citing turnips,
mustard, mint, and camomile as among the worst offenders. Several
of these "have overgrown the names of the valleys and imposed their*

own as in the case of Mint Valley on the seacoast, which was former-
ly called Rucma, and others." In Lima, endive and spinach grew taller
than a man, and "a horse could not force his way through them."

The most expansionistic European weed in 16th century Peru
was trebol, a clover or clovers that took over more of the cool, damp
country than any other colonizing species, providing good forage but
smothering crops as well. The former subjects of the Inca, who had
abruptly found themselves with a new elite and a new God to sup-
port, now discovered themselves in competition with trebol for crop
land. What was trebol? Most of it, in all likelihood, was white
clover, which performed the same role of pioneer and conquistador in
North America."[74]

Given such precedents, it is not difficult to imagine the results if
transgenic organisms created in the laboratory to "improve" agriculture should escape farmers' control and establish themselves independently. Not only would such plants and animals—in such cases,
perhaps more accurately called plant-animals or "plantimals"—have
no natural controls in their immediate environments, they would
have none in any environment, anywhere in the world.

The corporations which are developing such organisms, however,
are eager to introduce them and much effort is being expended to
overcome government constraints to their quick release. In the U.S.,
the Department of Agriculture (USDA) is favorable to such introduc-
tions, Goldburg reports:

In a highly controversial decision, the USDA in December 1994
allowed Asgrow Seed Company to sell squash genetically engineered to
resist two plant viruses. The engineered squash will undoubtedly
transfer its two acquired virus-resistance genes to wild squash
(Cucurbita pepo), which is native to the southern U.S., where it is an
agricultural weed. If the virus-resistance genes spread, newly disease-
resistant wild squash could become a hardier, more abundant weed.[75]

The general public—not just those who own or operate farms—has an obvious stake in preventing reckless introduction of laboratory-created organisms that could threaten not only our food supply, but the overall environment in which we live.

Also crucial to the environmental well-being of the general public is the system of regulation that guarantees the safety and purity of the foods we eat. The effects of the so-called General Agreement on Tariffs and Trade (GATT) and the North American Free Trade Agreement (NAFTA) have been to make food safety regulations the subject of dispute between GATT and NAFTA signatories, some of whom regard such rules as indirect barriers to trade.

In the past, food safety standards were generally within the jurisdiction of national, sometimes of provincial or state governments. Their strictness varied from country to country, with regulations in North America and Western Europe generally being more stringent than those of other nations. The new GATT regime, however, has imposed a single international standard: the Codex Alimentarius (Latin for "food code"), established and administered by a joint UN Food and Agriculture Organization (FAO) and World Health Organization (WHO) commission. Though Codex regulations are stricter than their national counterparts in many Third World and former East Bloc countries, many of the code's sections fall short of the standards of more technically advanced industrialized nations. Consumer groups in the north fear the Codex will be used to drag their national standards down, weakening protection against possibly dangerous food additives and other substances.

Stalin Redux: Collectivizing Rural America

6

IN THE EARLY 1930S, THE PARANOID, mass-murdering Russian dictator Joseph Stalin got it into his head that his nation's most successful farmers—the *kulaks*, or rural proprietors—were enemies of the Bolshevik Revolution. The communist ideology called for the organization of all of society on an industrial mass-production basis, and he was convinced this revolutionary reorganization should include agriculture.

The kulaks, whose deep understanding of the land and canny mastery of local and regional farm products marketing had helped them prosper for decades following the formal abolishment of serfdom in Russia in 1861, resisted Stalin's efforts to "collectivize" farming.[1] And the dictator's reaction was typically brutal. Backed by his army and secret police, he launched a purge not only of the entire kulak class, but of all of Russia's peasant family farmers. Between 1932–33 thousands were arrested and shot on the spot, or shipped off to exile in Siberia. Crops were confiscated without compensation, or simply burned in the fields, leading to a mass famine that boosted the death toll into the millions. Before this one-man disaster ended, more than 10 million individual Russian farm people were killed.

Russian agriculture never recovered.

In two years, Stalin had wiped out his country's rural brain trust, eliminating the very people whose comprehensive knowledge of crops, weather, and soil had assured the former Russian Empire of the Tsars a firm agricultural foundation. Now that skilled human foundation was gone, replaced by gangs of untrained field workers drudging under a variety of appointed Communist Party hacks, with no personal investment in the land they were assigned to cultivate, trying to force the natural world into a crazy pseudo-industrial model that had little connection with reality. The bumper crops that had once made regions like the plains of the Ukraine the wheat breadbasket of Eastern Europe, were no more. For almost the entire 70-year history of the Soviet Union, agricultural production declined. By the time the Cold War came to its inexorable end under Gorbachev in 1991, Russia had become heavily dependent on imports from Canada and the United States for its basic supply of grain.

Even today, more than 14 years after the collapse of communism, Russian agriculture remains an international basket case.

REPEATING HISTORY, SLOWLY

Since roughly the end of the Second World War, a similar, if somewhat less physically brutal, process has been slowly playing itself out in North America, with roughly similar results. America's family farmers—our equivalent of the kulaks—have been gradually wiped out in favor of an imposed industrial model, as divorced from the reality of the natural world as the giant collective farms of the defunct Soviet Empire.

North American agriculture has been—and is still being—collectivized. Only this time a semi-crazed Bolshevik dictator isn't behind it. The authors of this purge are the CEOs of the continent's largest agribusiness corporations, and the national governments their lobbyists and political campaign fund managers effectively control.

When the misguided policies of the Soviets brought agricultural disaster, they had export crops from North America's farmers to turn to and rescue them—at a price—from their own folly. But if North America imitates their sad history, who will rescue us? And who will rescue the natural environment, not to mention what little is left of the shredded social fabric of our rural communities?

As already noted in Chapter Four, the rural populations of the U.S. and Canada once included the vast majority of citizens, but by 1993 less than 2 percent of North Americans were farmers. The number has continued to decline, not only here but around the world.

Mechanization, capitalization, and consolidation of smaller farms into bigger ones—a growing number of which are either corporate-owned or locked into supply contracts with corporate buyers—have so "rationalized labor" on the Chicago School economic model as to wipe out the rural populations of most of North America, large parts of Europe, and Japan. The statistics are inescapable.

For example, a decades-long trend of rural out-migration in Canada became particularly intense after the Second World War. Between 1950 and 1980, the number of people living on farms "was slashed by a full 50 percent. In Ontario alone, nearly 362,000 people left the land—the equivalent in urban terms of the entire population of an industrial city the size of Hamilton—suburbs included—packing up and walking away from their homes."[2]

In the U.S., during the five-year period between 1981–86 alone, "the number of American farms had dropped by 15 percent,"[3] and of those that remained, the larger corporate-owned or corporate-dominated operations accounted for a greater and greater percentage of farm sales and revenues.

In Canada in 1992, only one-third of the nation's farms accounted for 79 percent of total farm revenue.[4] The statistics for the Canadian prairie province of Saskatchewan—whose grain has for decades fed not only Canada but much of the world, and whose rich soils make it ideal for farming—are a telling example of what is happening. Between the mid-1930s and 1991, the number of individual farms in Saskatchewan fell by 42 percent.[5] The average farm size, about 360 acres in 1918, had ballooned to more than 1,000 acres by 1991—an increase of more than 275 percent. By 1993, the Farm Credit Corporation owned one out of every 47 acres cropped in the province, as a result of farm foreclosures.[6] If families cannot farm successfully in such a fertile place as Saskatchewan, with so many worldwide markets, where can they?

As noted earlier in the case of California's Mexican farm workers, mechanization has accelerated the process of consolidation, not only by throwing farm laborers out of work (50,000 in the case of the *braceros*), but by assuring that only the larger, single-crop farms survive. Smaller family farmers in California, who produced a mixture of crops, could not afford the price of the new mechanical tomato harvesters, used for only a relatively small part of the year, and had to sell out to the big, single-crop operations that could afford them, slashing the number of California tomato producers from 4,000 to a pathetic 600.

Some of those remaining California "farmers" are actually multinational corporations. And those who are not may have very little real independence or decision-making power left. Working land that may be heavily mortgaged, saddled with further debt from the purchase of machinery and inputs, they are often locked into ironbound contracts with food processors. The latter provide them with the processor's own proprietary seed varieties and may dictate everything from fertilizer, irrigation, and herbicide regimes to the time and method of harvest. Nominally, these are self-employed family farmers, but in fact they function as little more than financial serfs, taking their orders from the processing giants and reducing the art of farming to something akin to an outdoor paint-by-numbers hobby kit, whose products are judged not by consumers, but by corporate accountants cubicled in some distant city high-rise.

According to the United Nations Food and Agriculture Organization, similar demographic movements have occurred in France, Germany, and Japan, with significant drops in agricultural population between 1961 and 1993. Overall, the European Union has seen its agricultural population—already considerably reduced from pre-Second World War levels—drop by 14 percent between 1961 and 1993.

According to the FAO, the agricultural population of the Soviet Union, including Russia, also fell steadily from 1963 to 1984, as farming became increasingly mechanized and the size of state and collective farms grew. It has kept falling in most successor states since the destruction of the Soviet empire. In many Third World countries, only massive growth in the general population has prevented a similar drop in rural numbers. Migration from the impoverished countryside to cities proceeds at an unsustainable rate, causing acute urban environmental and employment problems, but the rural population continues to grow faster than its members can flee to the towns.[7] In Kenya, for example, total population grew from 8,592,000 in 1961 to 26,090,000 in 1993—an astonishing increase of 303 percent—while agricultural population jumped from 7,473,000 to 19,737,000 in the same period—a 264 percent rise.[8] The country's rural economy, crippled by falling world prices for such commodities as coffee, tea, and sugar, and the unfavorable terms of trade created by the recent Uruguay Round negotiations under the General Agreement on Tariffs and Trade (GATT), cannot provide for its newest arrivals—nor can the overstrained urban environments of Nairobi and Mombasa to which so many migrate.

Throughout the twentieth century, all over the world, agriculture has shed labor in massive numbers, first depopulating the countrysides of most northern nations, and now beginning to drain those of the global South and to complete the process of concentration—already well-launched by communism's state and collective farm systems—in the poorer parts of the former East Bloc. As *Washington Post* reporter Nick Kotz writes, observing the U.S. scene:

> *The medium to large-size family farms—annual sales of $20,000 to $500,000—survived earlier industrial and scientific revolutions in agriculture. They now face a financial revolution in which the traditional functions of the food supply system are being reshuffled, com-*

bined, and coordinated by corporate giants. "Farming is moving with full speed toward becoming part of an integrated market-production system," says Eric Thor, an outspoken farm economist. "This system, once it is developed, will be the same as industrialized systems in other U.S. industries." Twenty large corporations now control [all of U.S.] poultry production.[9]

Describing the entry of oil and chemical companies, including the giant conglomerate Tenneco Inc., into farming, Kotz asks: "Will agriculture become—like steel, autos, and chemicals—an industry dominated by giant conglomerate corporations like Tenneco? In that case the nation will have lost its prized Jeffersonian ideal, praised in myth and song, of the yeoman farmer and independent landowner as the backbone of America."[10] The industrialization of agriculture, he writes, has further serious implications:

1. The future shape of the American landscape. Already in this country, 74 percent of the population lives on only one percent of the land. If present trends continue, only 12 percent of the American people will live in communities of less than 100,000 by the 21st century; 60 percent will be living in four huge megalopoli, and 28 percent will be in other large cities;

2. The further erosion of rural life, already seriously undermined by urban migration. Today 800,000 people a year are migrating from the countryside to the cities. Between 1960 and 1970 more than half our rural counties suffered population declines. One result is the aggravation of urban pathology—congestion, pollution, welfare problems, crime, the whole catalog of city ills;

3. The domination of what is left of rural America by agribusiness corporations. This is not only increasing the amount of productive land in the hands of the few, but is also accelerating the migration patterns of recent decades and raising the specter of a kind of twentieth-century agricultural feudalism in the culture that remains.[11]

Nor are those families being pushed off the land simply "ineffi-cient" or "behind the times" in their methods, as the Toronto *Globe and Mail* showed in a 1996 report on the closing of a Kabeka, Ontario, dairy farm:

> *"Well," said my neighbor, settling heavily into her chair, "that's it; we've sold the cows."*
>
> *The rest of the ladies' Scrabble group took a minute to absorb the news. Sold the cows? No more dairy farming?...*
>
> *The news was a shock to us all. It was one of those times when we didn't know what to say. They'd been doing well at it, well enough to support her and her husband, their four sons, his parents and his younger brother. Two houses, two barns, two farms, four silos, 50 cows, a flock of chickens, hay fields rented from neighbors— it was quite an operation. Of the five dairy farms left in our town-ship, I'd have said theirs was the best.*
>
> *The farm first belonged to his parents. When they came here from Holland in the early 50s, they took over a farm that had been neglected and made it into a wonderful example of an efficient dairy operation. It was something to be proud of. The farm is one of the oldest in the township, having been started in about 1895, when the first settlers were trying to make a go of it in this inhospitable region of Ontario.*[12]

The article continues:

> *"We're young enough to do something else," my neighbor explained. "If we want to stay in farming, we'd have to expand so much that we'd be in debt for 20 years. We'd be lucky to have it paid off by the time we retire. We'd be working as hard as we work now but it would all go to paying off debt. We'd need to have more cows, more hay to feed the extra cows, more expenses for new regulations—it's too much."*

I think she's still in shock. I know I am. It's sad to see the end of something, and maybe sadder still to know that if they can't make a go of it, neither will the four farms still operating here. Those four are just postponing the inevitable.

Next year, changes resulting from the free trade agreement will mean that Canadian dairy farmers will no longer enjoy the protection of any sort of regulation. We will see more of our milk and dairy products coming from the U.S., and eventually, when I am in a gloomy mood, all of North America will be served by three dairy farms, each the size of Nova Scotia.[13]

The article author's fears may be less exaggerated than she realizes. According to another *Globe and Mail* article, published on page one on April 30, 1996, as a result of the North American Free Trade Agreement (NAFTA) the existence of some 32,000 Canadian farms could be endangered.[14]

Endangered along with individual farms are the small- and medium-sized towns those vanishing farm families once supported. As the countryside slowly empties, the towns, one after the other, repeat the same experiences.

First, the local school district begins to consolidate grade schools, and then high schools. The smaller schools, both public and parochial, go first, and the morning and afternoon rides on the yellow schoolbus get that much longer for rural kids. Before their families at last give up and leave the land, some kids may spend as much as four hours on the bus each and every day, traveling to and from their rural district's ever-more-crowded consolidated schools. After the schools go local restaurants, and the Greyhound or Interprovincial bus stops they usually host. Then goes the local post office, whose boxes are merged with those in a larger town farther away, while rural mail delivery is curtailed.

Next go the hardwares, groceries, and general stores, forced to close for lack of clientele. The farm families that remain must now spend hours on Saturdays driving to a distant chain store in a distant town to get their coffee and sugar and shoes for the kids.

On the prairies, the grain elevators close down, and branch rail lines are shut down. Less money is devoted by state or provincial governments to road repair, and the local routes start to erupt in potholes, like a child with measles. Then finally, the churches start to consolidate. At first, the building where dad and mom were married and in whose yard grandpa and grandma are buried, gets a weekly or biweekly visit from the pastor of a church in some neighboring town. Then, as the neighboring towns start to die, the church is boarded up, and anyone who wants a Sunday service must be willing to spend half the Lord's day at the wheel, driving down the potholed road to hear God's word.

At last, the towns become ghost towns, or near to it. In Saskatchewan, some desperate rural municipalities have taken to advertising empty homes and lots for sale for a dollar. The homes were abandoned when their departing owners found no buyers, or were expropriated for nonpayment of property taxes, and the municipalities are hoping the price will lure retirees. But what aging pensioner will move to a town that has no doctor? The doctors, by this time, are gone too.

The process—as this author knows from personal experience—is one of the most depressing things a person who loves the land can witness.

Of course, the same trends the *Washington Post*'s Kotz and the *Globe and Mail* deplore are working—or already have worked—the same kinds of changes in the industrialized economies of Western Europe and Japan. Bookchin laments the increasingly common worldwide result:

Agriculture, in effect, differs no more from any branch of industry than does steelmaking or automobile production. In this impersonal domain of food production, it is not surprising to find a "farmer" often turns out to be an airplane pilot who dusts crops with pesticides, a chemist who treats soil as a lifeless repository for inorganic compounds, an operator of immense agricultural machines who is more familiar with engines than botany, and, perhaps most decisively, a financier whose knowledge of land may beggar that of an urban cab driver. Food, in turn, reaches the consumer in containers and in forms so modified and denatured as to bear scant resemblance to the original. In the modern, glistening supermarket, the buyer walks dreamily through a spectacle of packaged materials in which the pictures of plants, meat, and dairy foods replace the life-forms from which they are derived. The fetish assumes the form of the real phenomenon. Here, the individual's relationship to one of the most intimate of natural experiences—the nutriments indispensable to life—is divorced from its roots in the totality of nature. This denatured outlook stands sharply at odds with an earlier animistic sensibility that viewed land as an inalienable, almost sacred domain, food cultivation as a spiritual activity, and food consumption as a hallowed social ritual.[15]

American poet and social critic Wendell Berry, himself a farmer, has also identified and warned against this process of agricultural industrialization and rural depopulation, which he calls "a work of monstrous ignorance and irresponsibility on the part of the experts and politicians, who have prescribed, encouraged, and applauded the disintegration of farming communities all over the country."[16] Like Kotz and Bookchin, Berry sees links between the rural and urban crises, and fears the cultural effects:

Few people whose testimony would have mattered have seen the connection between the "modernization" of agricultural techniques and the disintegration of the culture and the communities of farming— and the consequent disintegration of the structures of urban life. What we have called agricultural progress has, in fact, involved the forcible displacement of millions of people.

I remember, during the 50s, the outrage with which our political leaders spoke of the forced removal of the populations of villages in communist countries. I also remember that at the same time, in Washington, the word on farming was "Get big or get out" a policy which is still in effect and which has taken an enormous toll. The only difference is that of method: The force used by the communists was military; with us, it has been economic—a "free market" in which the freest were the richest. The attitudes are equally cruel, and I believe that the results will prove equally damaging, not just to the concerns and values of the human spirit, but to the practicalities of survival. The aim of bigness implies not one aim that is not socially and culturally destructive.[17]

Berry insists food is "a cultural product; it cannot be produced by technology alone"—that is, not unless the process is radically simplified, as it is in highly mechanized, industrial mono-cropping (single-crop) systems. As described in the previous chapter, massive acreages are leveled and sown, year-after-year, with no or only infrequent crop rotations, to a lone, high cash-return crop such as hybrid corn, which quickly depletes soil nutrients. To make up for the lost nutrients, especially nitrogen, heavy doses of inorganic chemical fertilizers are employed, which "burn" living soil organisms and pollute the water table. The industrial division of labor involved in such environmentally destructive "factory farming" also multiplies the number of wage-worker "specialists" doing the work, each focused on his narrowly defined task, while eliminating generalists who, like the vanished kulaks and vanishing American family farmer, can envision whole systems.

Like Williams, Berry sees this as symbolic of a social outlook that now runs through virtually every aspect of life, one that favors compartmentalization and leads to "a radical simplification of mind and character":

That the discipline of agriculture should have been so divorced from other disciplines has its immediate cause in the compartmental structure of the universities, in which complementary, mutually sustaining and enriching disciplines are divided, according to "professions," into fragmented, one-eyed specialties. It is suggested that farming shall be the responsibility only of the college of agriculture, that law shall be in the sole charge of the professors of law, that morality shall be taken care of by the philosophy department, reading by the English department, and so on. The same, of course, is true of government, which has become another way of institutionalizing the same fragmentation. However, if we conceive of culture as one body, which it is, we see that all of its disciplines are everybody's business. [18]

The "compartmental" mind-set, symbolized by the factory-farm, is symptomatic of a culture of alienation, whose components are cut off from one another as well as from nature. But, Berry asserts, "a culture cannot survive long at the expense of either its agricultural or its natural sources. To live at the expense of the source of life is obviously suicidal."[19] By way of example, he points to the comments of former U.S. Secretary of Agriculture Earl Butz, and former Secretary of Defense Robert McNamara, chief architect of the Vietnam War:

Our recent secretary of agriculture remarked that "Food is a weapon." This was given a fearful symmetry indeed when, in discussing the possible use of nuclear weapons, a secretary of defense spoke of "palatable" levels of devastation. Consider the associations that have since ancient times clustered around the idea of food— associations of mutual care, generosity, neighborliness, festivity, communal joy, religious ceremony—and you will see that these two secretaries represent a cultural catastrophe. The concerns of farming and those of war, once thought to be diametrically opposed, have become identical.[20]

If what such critics claim is true, a kind of circle has been closed. Farming, the basis of settled human life, has permitted the development of a civilization which alienates and cuts human beings off from life. By helping depopulate the land, to overcrowd and overwhelm the cities and degrade both the quality of our food and of the natural environment, the farming systems developed by our culture run the risk of destroying themselves—and thus the foundation of our civilization.

"This is not merely history," writes Berry. "It is a parable."[21]

NO ACCIDENT

The causes of this ongoing disaster are not accidental, nor simply the inevitable result of some blindly fated "march of progress" brought on by technological development. Although technology (such as the invention of mechanical tomato harvesters) has played its part, they are chiefly the result of an overall macroeconomic policy, which has been determined by the agendas and profit margins of greedy speculators, of the hegemonic ambitions of the largest agribusiness corporations and banks, and of the tax, international trade, and farm support policy decisions of national governments, especially the U.S. government.

The mechanism which has accomplished their mutual goal of eliminating the family farm and "rationalizing" food production on an industrial, communist-collective-lookalike model is often referred to as the "cost-price squeeze." And this squeeze is without mercy. As the author of this text put it in an earlier report:

> Today's average [food industry] corporation is not only multinational and multi-industry (horizontal integration), but involved within each industry in the entire production chain from basic supplies to product marketing (vertically integrated). Needless to say, the individual farm operator—who according to Revenue Canada in 1979 boasted an average net income of only $12,598—has become an

almost insignificant economic cipher, overwhelmed by the financial might of those with whom he must deal.

And yet he has no choice but to deal. On the one hand he must buy equipment and supplies—machinery, fuel, feed, seed, and fertilizer—which along with rocketing land prices and the interest on his loans make up the farmer's cost factor. On the other hand, once a crop is raised it must be sold to buyers in the processing, distribution, and retail (PDR) sector who influence the farm gate price the producer receives—the second factor in the squeeze. Obviously, the contest is unequal at both ends of the production cycle.[22]

Sometimes, particularly in the U.S., the same corporate conglomerate faces the farmer at both ends of the cycle, selling him his seed, fertilizer, herbicides, and pesticides, then buying his end product, processing it, and marketing it to the continental supermarket chains—which may have some of the same directors sitting on their interlocked corporate boards. The district supermarket to which struggling farm families must drive on Saturdays after their local town's independent stores have disappeared, may ironically be owned by one of these very chains. Insult added to injury.

Until recently, Canadian farmers had a degree of bargaining power in this struggle thanks to the mediating function of the country's marketing boards. Marketing boards group individual producers of a given farm commodity, such as eggs, milk, or wheat, who pool their production and agree to sell at the same price. This provides a countervailing weight to that of the corporations and helps the board's farmer-members to ease the cost-price squeeze that would otherwise put many of them out of business. In effect:

Many boards serve, in part, as farmers' unions. Federal board officials are usually government-appointed, but most provincial and local boards are elected by their farmer-members, just as the officers of a union local would be. They negotiate, sometimes demand, a return on farmers' labor in much the same way that a union bargains for workers' wages.[23]

This bargaining ability, of course, runs counter to the interests of the corporations with which boards deal. Economic self-interest should impel the corporations to try to neutralize or eliminate marketing boards. Farm writer Terry Pugh succinctly summarizes the underlying agenda of the General Agreement on Tariffs and Trade (GATT), whose international bargaining rounds in the 1990s led to the formation of the World Trade Organization (WTO). According to Pugh, the significance of the presence of Cargill's Whitney MacMillan at the trade talks became self-evident:

> *In the grain sector, for example, a half-dozen corporations influence global prices and supplies, and design trade policies which accommodate their self-interest. The largest of this oligopoly is Cargill Grain, closely followed by Archer Daniels Midland (ADM), Continental, Louis Dreyfus, Bunge & Borne, Mitsui and Feruzzi. These companies' overriding need to source raw materials from the cheapest suppliers, and acquire unfettered access to expanding markets, provided the initial impetus and ongoing direction for GATT negotiations. Domestic farm programs, including Canada's orderly marketing and supply-management systems, represent obstacles to this flow of commodities and capital. Consequently, marketing boards were targeted for elimination in the trade talks.*[24]

But small family farmers were targeted for elimination on a national basis long before these international negotiations, by figures like former U.S. Secretary of Agriculture Earl "Get big or get out" Butz and former Republican U.S. President Richard "I am not a crook" Nixon, of Watergate fame. Butz, a former board member "of such agribusiness giants as Ralston Purina, International Minerals and Chemical Corp., Stokely Van Kamp, Inc., J.I. Case & Co. and Standard

Life Insurance Company of Indiana,"[25] served more than one U.S. president, but is best known for his tenure in the agriculture portfolio under Nixon.

As researcher A. V. Krebs explains, Nixon had decided early on to "rationalize" American agriculture along the lines of the economic theory of "comparative advantage" (previously described in Chapter Four). He based his planning on a 1972 report by a Presidential Commission on International Trade and Investment Policy, headed not by any farmer, but by International Business Machines (IBM) finance executive Albert L. Williams. As Krebs outlines it:

> *The Williams Commission believed that the United States had a nat-*
> *ural advantage in grain production due to highly favorable soil and*
> *weather conditions combined with intensive application of technol-*
> *ogy and capital, thereby making it a model of "capitalist efficiency."*
>
> *To carry out such "free trade" policies, U.S. agriculture would*
> *have to be converted into an efficient export industry, phasing out*
> *domestic farm programs designed to protect farm income and mov-*
> *ing to a "free market" oriented agriculture. This approach was wide-*
> *ly supported by corporate agribusiness and would become the corner-*
> *stone of the Nixon administration's farm policy.*[26]

The idea was, basically, to convert the countryside into a vast, mass-producing grain factory—and later, under succeeding presidents, into factories manufacturing tomatoes, oranges or whatever else. The bigger–and thus more theoretically "efficient"—these outdoor factories were, the better. And Butz applied himself enthusiastically to the task, urging farmers "to plant 'fence row to fence row' ('go borrow, young man, bet bigger')."

Krebs continues:

> *By structuring USDA [U.S. Department of Agriculture] programs to*
> *support the Nixon doctrine of using agricultural products to stabi-*

*lize the nation's balance of payments, [Butz] encouraged land spec-
ulation, and by preaching that "planting fence row to fence row" was
almost a sacred obligation, Butz must now bear a major responsibility for the plight of thousands and thousands of American farmers
who have been forced off their land and into bankruptcy, forced foreclosures, depression, divorce, alcoholism, and suicide.* [27]

Nixon's policies have been continued, with a few detours and
modifications here and there, by most U.S. administrations since,
especially that of Ronald Reagan. And the so-called Uruguay Round
of GATT negotiations of the 1990s, which led to the establishment of
the World Trade Organization, adopted a similar, "global rationaliza-
tion" philosophy toward agriculture.

For example, several elements of the final agreement's import
protection regime affected farming, but the most important dealt
with tariffication, a process which converts all presumed import barriers—even indirect ones—into quantifiable tariffs. A United Nations
Food and Agriculture Organization (FAO) report summarizes:

*The new rules required that all quotas, variable levies, and other
import barriers be converted to common tariffs, as soon as the agreement took effect. These and existing tariffs had then to be reduced by
a minimum of 15 percent each over the implementation period with
the tariff reductions as a whole having to average 36 percent.
Developing countries were required to reduce tariffs by 24 percent
and were allowed 10 years, instead of 6, to implement the cuts. The
tariffication of non-tariff barriers and the prohibition against future
use of such non-tariff instruments represents a major reform of the
trade rules affecting agriculture. It should bring transparency to barriers that have been hidden from public view and should also expose
the high levels of protection enjoyed by agricultural producers in
some countries.* [28]

Most commentary on the agreement, particularly by spokesmen representing the conservative, corporate agribusiness view, applauded the new, open-borders regime. For example, economist Stefan Tangermann of the University of Göttingen, Germany, was quoted widely to the effect that the Agreement on Agriculture was "a historical breakthrough ... a major step in a good direction."[29]

Nor would the barriers at international borders be the only ones to come tumbling. Calling attention to a feature overlooked by many other observers, Tangermann added:

> *Where protection against imports can no longer be provided at will and export subsidies will have to be cut back, there is the danger that domestic subsidies will be used instead. It is therefore good to know that there are now also limits to the extent to which governments can provide domestic support."*[30]

The assumption of Tangermann, and spokesmen for most G-7 countries (France was a notable exception) whose trade representatives dominated the GATT bargaining, was that removing protection for domestic farm industries would—automatically—improve the previous system.

The question "improvement for whom?" went unspoken, except by a minority of critics like British economist/authors Tim Lang and Colin Hines, whose book, *The New Protectionism*, attacks free-trade assumptions.[31] Lang and Hines were blunt in a January 1996 article that followed publication of their book:

> *The new GATT has enshrined the 1970s Reagan/Thatcher era formulae: deregulation, economic efficiency, international competitiveness, and a dogmatic reliance on the market to meet all needs. Yet there is much evidence these policies don't work for the common good. The large farmer, traders, and big companies benefit, but the evidence is that intensive, high-input farming, the logical outcome of*

these policies, is disastrous for the environment, rural economies, food quality, and food security.[32]

That the new GATT/WTO system would threaten Canada's farm marketing boards system, while favoring the transnational corporate sector, was obvious. The boards were even more directly threatened by North American Free Trade Agreement (NAFTA) provisions, as Canadians discovered in 1995-96 when the U.S. filed a brief under NAFTA terms demanding that Canada quash border tariffs designed to protect farmer-members of Canada's dairy, egg, chicken, and turkey marketing agencies.[33]

DARK, SATANIC BARNS

The corporate, factory-farm food system that dominates so much of North American agriculture today is destructive of nearly everything it touches. It degrades the nutritional quality and taste of the food we eat, filling it with toxins and poisons, destroys family farmers and rural communities, blights the land and the environment, and tortures the living creatures it "manufactures" in its dark, satanic barns (to paraphrase William Blake). Its future products may prove to be genetic or micro-mechanical horrors, inflicting new plagues upon an unprepared world.

And the only entities who seem to truly benefit from this system are a tiny group of already-wildly-rich corporations and their executives, and those, including our politicians, who have been coopted by them. It is the product of their brains and their ideologies, not the result of any inexorable "invisible hand" of economic fate.

It is unscientific, unnatural, and for ordinary citizens ought to be unacceptable.

Part II : THE SOLUTION(S)

Each time we exhale: carbon
dioxide, life to green things!

Each time we inhale:
phyta in countless divisions,
all mosses, clinging
to high rocks, spined cacti,
halophytes in tide marshes,
great sequoias, thousand-
year/boardfoot/girth/thicknesses
of centimeter bark, despite
chainsaw raaazzzz,
despite falling, despite sawdust,
despite whine of circular blade,
grasswaves of prairie windtides,
swayback, springing back, scythed,
unscythed, old oaks erect like fossil
cocks, deep-rooted maples
thick with syrup, rattling
leaf of poplar, and
plankton gliding, floating on the
slipping, sliding, oilslick surface
of the sea—
send
oxygen.

This is the secret
of breathing.

IT WASN'T A BIG PLACE: BY TODAY'S STANDARDS,

barely a postage stamp. A few Holsteins and, if imperfect memory serves, a Jersey or two, shared the dairy barn where a shining-clean stainless steel milk storage tank received its daily gallons of product. There were chickens in a coop, a kitchen garden just outside the white frame farmhouse, fields of grain and pasture beyond, and somewhere in the middle, a pond with willow trees around it.

It was my cousins' mixed dairy farm near Chelsea, Michigan, and it gave me the gift of a true understanding of summer, of what a summer day could mean.

We were haying for around two days, hot days, out on the flat hay wagon, towed just behind the baler by what may have been an International Harvester, or was it a Cockshut tractor? No matter. Of course, in those days, it was all square bales; today's big round balers hadn't made an appearance yet in Michigan. They'd come trundling up the conveyor belt from the baler, one after the other, the hay bales weighing in at 80 pounds and the straw bales, when we did the wheat field, at around 60.

Taking turns, we'd hook one end of each bale with the clawlike, steel hay hooks in our right hands, grip one of the two cords the baler had just wrapped around the square bundle in our left hands, turn and toss the bale up on the pile behind us. There, a third person would catch the bales and stack them in an overlapping pattern, layer on layer, so they'd stay steady and not fall off. It was hot, sweaty work in the bright sun. Prickly work, too. By the time the day was over, forearms and thighs were itchy with tiny punctures made by the stems.

At noon, we took a break to eat the lunches my cousins' mom had packed for us, and maybe drink some cool, homemade cider, kept cool by a couple of wet towels wrapped around the jug, where it was tucked down between some tree roots in the shade of the fence row. As we sat and ate, we could watch the red-tailed hawks circling lazily over the fields, hoping to spot any field mice thrust suddenly into the open by the act of our passing by, gathering, and baling the mown crop that had previously hidden them. Every now and then, one of the hawks would tuck in its wings and dive, silently, straight down onto some hapless, scurrying prey.

Lunch over, it was back on the wagon again, to finish the field. And later in the afternoon, there was the bumpy ride back to the barn, where we'd set up the elevator and hoist bales off the wagon onto its clattering conveyor, then watch as they'd mount up to the loft. There one or two of us would be standing to catch them and, once again, stack them in patterns—future winter feed and bedding for the cows. It was hard work, and by the time we trooped in for dinner, we were genuinely tired.

But dinner was always satisfying, with everyone in a good mood that we'd gotten the hay in, especially if a summer shower had threatened and we'd had to race the rain.

I still recall one night, after one of those hot days, when we kids decided to go outside to the pond and camp out under the willows. At dusk, we set up the tent and laid our sleeping bags down. Then, as darkness fell, we lit a fire and toasted marshmallows, talking, joking, telling yarns. I told a "shaggy dog" story that night, out there under a half moon, with the embers glowing and my cousins sitting around listening as the story got wilder and sillier. And later, when we fell asleep in our sleeping bags, we slept the sleep of the just, soundly.

I've never forgotten those days, and for years afterward, in other times and places, the images associated with them have returned, even in—perhaps especially in—the city.

Like the summer of unemployment in Detroit. It was after I'd quit my editing job at a wire service bureau, whose bureau chief was a paranoid and whose staff members were a weird mix of boozers, pep pill addicts, and toadies. The character of the place was probably best exemplified by the president of its Wire Service Guild local, who had crossed the picket line in the middle of a strike and been rewarded by management for his act of scab betrayal with instant promotion to the post of news editor. One of my fellow rewrite men used to take notes on what bureau staff said or did during the day, and filed a secret, nightly snitch report to the paranoid bureau chief.

After a few months in this decidedly unhealthy environment, I found another job and quit the bureau, in such a way as to assure that I would never be hired back. Then the outfit that had offered me the new job reneged—and I found myself out of work, with a mortgage, car payments, and a family to support. It was not a pleasant period.

What kept me sane during that bleak summer was the memory of those earlier times in Chelsea, and a tiny, four-by-six-foot vegetable garden I planted next to our mortgaged house. I don't recall exactly what was in the garden, except for carrots and some beans, and it could have taken no more than a half hour a week to take care of it. But I spent much more time than that there. In fact, I went outside every morning before starting the daily job hunt, and looked at it, bending down to pull out any stray weeds, watering it, watching every minute change in each and every plant. I'd stand there, just looking at it, for 20 minutes, 40 minutes at a time until my wife thought I was cracking up.

But I wasn't cracking up. I was in contact with something real, something that steadied me and kept my thinking from veering off into total pessimism. In terms of produce, we may have gotten one or two plates of beans and a salad or two out of that little endeavor. But it gave me much more, incalculably more.

So did the tiny, flower-bed sized garden I've planted for the past two years in the yard outside my condominium in flat Saskatchewan. Under the rules of the condominium corporation in the development where I live, everyone has to keep their yard in plain grass, except for a specified area of "flower border." Rather than plant that border in flowers, I put it in vegetables—and managed bumper crops (for the minuscule square footage) two years running. The first year it was mostly tomatoes, and the second beans. Lacking the possibility of expanding outward beyond the borders, I went upward, installing a network of trellises and strings, along which the beans ran wild. The first year I harvested a good bushel and a half of tomatoes. The second year, I ended up with more than 20 quarts of canned beans, carrots, beet greens, and swiss chard, not to mention several fresh salads and plates of fresh beans.

And several wonderful, sunny mornings spent standing in the yard, just looking at green things growing.

I've owned and operated actual mixed farms of my own in Quebec and Ontario, raising grain and field crops and poultry and even trees (sugarbush and cedars for fencing). But these tiny gardens gave me almost as much as the farms did, in terms of morale. Gardens are like that.

Which, by a somewhat roundabout route, brings us to the subject of how to react to the industrial food system that has swallowed up so much of the continent and endangered our health and our food supply, not to mention our socio-economic relations and our natural environment. There are many responses to this threat, but the first and easiest—as well as the most immediately satisfying—is the simple act of planting a garden.

Given the present, corporate-dominated atmosphere in North America, doing so is very nearly an act of subversion. For thousands of years, in every unjust society, those who, from Robin Hood on down, have affirmed life over the version of reality promoted by the "powers that be," have been seen as outlaws. There is no reason why gardeners should be exceptions. Guerrilla gardeners. It has a nice, alliterative ring to it.

ANYWHERE THERE'S DIRT

And anyone, anywhere can plant a garden. Even living in the center of a big city, if you can find a spot where there is soil, dirt of any kind, living things can grow there, even in an apartment building where you are a renter. Flower pots on the balcony can be sown with vegetables instead of flowers, a patch of roof on the top of the building can be spotted with "wading pool gardens," planted in plastic children's wading pools (be sure, in this case, that the topsoil-filled pool won't weigh too much for the roof beams, and come crashing down into the living room of some hapless apartment renter below!).

From ground-floor flower borders or from balcony railings, strings can be run upward, balcony to balcony, to support climbing vines, pole beans, grapes and other species. In Montreal's famed "balconville," such climbing string gardens were once commonplace, not just among the French, but among Italian immigrant families, whose beans and tomatoes and grapes climbed across the face of their buildings, and whose basements were full of fermenting, homemade wine.

Urban and suburban homeowners, with the spatial luxury of their own backyards, have it better, with far more room to grow in. They can even plant trees—apples, crabapples, plums, or pears.

In contrast, people living in condominium developments, where rules on outdoor landscaping can sometimes be strict, may have to use a bit of ingenuity to maximize gardening space. But where there's a will, there's a way. If a condo actually forbids gardening (thankfully, a rare situation), chat up your neighbors and bring a motion at the next condominium board meeting to change the rules. Or run for a board seat yourself.

Once a spot to plant is found, any seeds, and any growing method, will be an improvement over supermarket produce, if only because the home garden's yield will be fresh, rather than frozen and trucked thousands of miles, then artificially ripened to look cosmetically good on a chain store shelf. But using organic growing methods, and choosing heritage rather than mass-market seeds, will bring a triple advantage: in food quality, flavor, and choice of varieties.

Inexperienced gardeners, and many city dwellers, are sometimes unsure what the term "organic" means, and regard organic gardening as some sort of mysterious, esoteric specialty accessible only to experts. The exact opposite is true. Organic growing is the oldest and simplest form of cultivation, the way all crops were produced for thousands of years. Modern, chemical/industrial growing methods, in contrast, are a recent departure from that norm, only a few decades old, and more expensive and complex than doing things the natural way.

Basic instructions for organic gardening can be found on the shelves of any bookstore or public library, with a variety of titles for beginning, intermediate, and expert home horticulturists. The best of all books on the subject, with one of the longest pedigrees, is *The Encyclopedia of Organic Gardening*, updated and published annually by the Rodale organization of Emmaus, Pennsylvania.[1] Its pages contain authoritative advice and step-by-step instructions on everything from composting to companion planting and non-chemical pest control techniques. My own hardbound copy is dog-eared, smudged, underlined, and margin-noted in so many different ballpoint pen inks and pencil markings, with so many reminder notes and newspaper clippings tucked between its pages, that it looks like a mini-library in itself.

In the past, I've also found the general, regional weather and planting data contained in annual almanacs such as *The Old Farmers' Almanac* (for the U.S.)[2] and the *Canadian Farmers' Almanac*,[3] useful and often accurate. But the predictions of most traditional almanacs are based in large part on averages of the recorded weather and climate data of past years. Climate change, brought about by the pollution-induced global warming phenomenon, has already begun to cause unpredictable shifts in what used to be normal weather patterns, and these fluctuations are likely to get worse as the planet's atmosphere struggles to find a new equilibrium. Rather than count on the unpredictable, I read almanacs now chiefly for their entertainment value.

Armed with the Rodale's advice, my backyard composter (purchased at the Canadian Tire store at a September end-of-season sale for $39.99), and a supply of non-toxic diatomacious earth "bug dust," I do battle every year with cabbage worms, slugs, and other competitors, and, most of the time, win. Or at least earn a draw.

The seeds for my garden, whenever possible, are not purchased from any of the big, corporate-owned seed houses that are slowly establishing monopoly, or at least oligopoly control over the conventional market. Instead, I purchase seed via mail order from a widespread network of heritage seed suppliers and seed-saver groups which have sprung up (so to speak) in reaction to the international seed crisis.

Few urban dwellers are aware of this particular crisis, but they ought to be.

As explained in Chapter Five, industrial agriculture tends to favor simplicity and uniformity over variety and diversity. If a single, thick-walled, hard-as-rock, flavorless, but uniformly sized and colored variety of tomato that always ripens at the same time is available, why bother with all those hundreds of other varieties? The fact that the others may be nutritionally superior or better-flavored, or may simply represent a relief from the monotony of eating the same thing over and over, day after day, is beside the corporate accountants' point.

And corporate agribusiness likes the idea of both horizontal and vertical economic integration—that is, the idea of owning most of the companies active in a single area, such as farm machinery (horizontal integration), as well as companies active at other stages of the crop production system, from seed sales to the manufacture of fertilizer, pesticides, and herbicides, to food transport (vertical integration).

The worldwide consolidation of seed production and sales into the hands of fewer and fewer companies has taken place in tandem with a record of blatant, irresponsible waste of genetic resources. One of the worst examples of the latter was the summer 2000 decision by Seminis, at the time the world's largest vegetable seed corporation, to "eliminate 2,000 varieties——or 25 percent of its total product line—as a cost-cutting measure."[4] The Rural Advancement Foundation explained the situation in a news release:

> *Seminis, a subsidiary of the Mexican conglomerate Savia, controls nearly one-fifth of the worldwide fruit and vegetable seed market and is the source of approximately 40 percent of all vegetable seeds sold in the United States. The company built its seed empire by acquiring a dozen or so seed companies, most notably the garden and seed division of Asgrow, Petroseed and Royal Sluis. As a result of its buying binge, Seminis offerings grew to approximately 8,000 varieties in 60 species of fruits and vegetables. On June 28, 2000, Seminis announced that it would eliminate 2,000 varieties—or 25 percent of its varieties—as part of a "global restructuring and optimization plan."*
>
> *No one knows for sure which varieties will be dropped from Seminis' commercial line, but the older, less-profitable, open-pollinated varieties will be the first to go. Seed corporations favor hybrids because profit margins are greater, because gardeners and farmers can't save hybrid seed (thus encouraging repeat customers), and because newer varieties are more likely to be patented or protected by plant variety protection laws.[5]*

To deliberately throw away, and cause to be lost forever, thousands of unique plants whose possible value to future generations may now never be known, is worse than irresponsible. It is an environmental sin. Many of those varieties may have contained the key to preventing new crop diseases, or to providing as yet undiscovered phytochemical products that could be used in human medicine, to cure human diseases.

Some of those lost varieties may have been used to develop plant traits that could respond to new climatic or soil conditions that we may soon face as a result of pollution and global warming, or may have contained compounds that could be used to create new chemical or industrial products. We will never know, now, what treasures may have been there. They are gone, for the sake of a slightly higher, quarterly profit margin in some cubicled accountant's ledger.

Of course it's possible that some of the discontinued varieties may still be preserved somewhere in a Seminis seed bank, or in a government or other laboratory, and thus may not be wholly lost. None of the news stories on the Seminis decision mentioned this. But whether a particular company keeps a tiny supply of plant germplasm in storage or not, those varieties are no longer available to growers.

SEED SAVERS

Countering this trend are a host of citizens groups, home gardeners and organic farmers who are part of a burgeoning movement to preserve traditional, non-gene-altered and/or rare plant varieties. Fifteen years ago, there were only a few such groups in North America, most notably the Seed Savers Exchange, founded by Kent and Diane Whealy of Decorah, Iowa, and the Heritage Seed Programme (now Seeds of Diversity Canada), headquartered in Toronto, Ontario.

Today, even a cursory Internet search with the Google search engine will turn up hundreds of groups, in nearly every country in the world. Two of the best websites on the subject are:

Seedsaving and Seedsavers' Resources:
http://homepage.tinet.ie/~merlyn/seedsaving.html
which includes links to companies, groups and individuals who trade, buy or sell heirloom seeds, as well as to resources such as tools, books, and other supplies. Sources are conveniently set off by national flag, to indicate where each one is located and in what language their material is presented

Seeds of Diversity Canada Resource List:
http://www.seeds.ca/rl/rl.php>
which contains a long list of mail order sources of heritage
and organic seeds. The list, and the Seeds of Diversity homepage, are posted in both English and French versions.

The Seed Savers Exchange publishes one of the best how-to books
on the techniques of saving seeds, Suzanne Ashworth's *Seed to Seed:
Seed Saving and Growing Techniques for Vegetable Gardeners.*[7] Written in
layman's language, Ashworth's book covers everything from seed
cleaning and storage methods to pollination and maintaining varietal
purity. She gives detailed directions for each of eight major vegetable
families, and the best known species in each family. I haven't seen a
better treatment of the subject.

Reading over any of the many mail order catalogues available is
one of the great late-winter/early-spring armchair gardener's pleas-
ures. Who can resist daydreaming after reading a description, in the
Southern Exposure Seed Exchange Catalogue and Garden Guide, of such
tomato varieties as:

*GERMAN RED STRAWBERRY: Distinctive in appearance, excels
in flavor. This German heirloom resembles a strawberry in color and
shape, though much larger! German Red Strawberry is excellent in
salads and is the quintessential sandwich tomato. It is meaty with a
scant amount of seed and juice, and the spectrum of flavor components runs broad and deep. Fruits average 10 oz. in weight, 3" wide
by 3 1/2" long, somewhat variable in size but fairly uniform in shape
producing an average of 2 fruits per cluster.*[8]

Looking at the photos in the *Seeds of Change Annual Seed Book*,[9] of
carrots, sweet corn, pole beans, garlic, and fat, round melons, is
mouth watering.

Before letting themselves get carried away, however, gardeners should keep in mind that the varieties they see in a catalogue from Santa Fe, New Mexico, may not do so well on the windswept prairies of northern Saskatchewan, or in a Laurentian Mountain valley in Quebec. The further north one goes, the shorter the growing season, with first and last frost dates that may be as much as two months earlier or later than in southern locations. Soils too can be vastly different from region to region, depending on their geological origins. Before ordering a supply of spring seed, first check the climate zone and soil maps on the U.S. Department of Agriculture and Agriculture and Agri-Food Canada websites, then read the seed catalogue descriptions closely, to see if "the fruit fits the field."

For a balcony garden using flowerpots, the only tool needed is a small hand trowel for scooping dirt into the pots from a bag of topsoil purchased at the local hardware or garden shop. And maybe some gloves. For a backyard garden, a spade and rake are all that's needed. If it's a bigger yard, a hoe might come in handy for keeping the weeds down between rows.

Most seed packets come with planting directions printed on the package back, detailing best planting dates, the depth at which to bury the seeds, and the number of days from planting to first harvest. The best protection from insect pests, in my own experience, is diatomacious earth. This is a white powder made from a ground-up mineral produced by microscopic creatures called diatoms. It is non-toxic to humans, but the tiny, sharp edges of the diatom-formed grains act upon insects much like razor wire does on people. I've used it to rid my garden of cabbage worms, slugs, and several other common pests. It's available in most hardware stores and garden shops.

You may also want to buy a lawn chair, to sit in while sipping a cold beer and watching your plants grow.

In early- to mid-season, fresh produce and fresh salads are the gardener's reward. In northern regions where winter is a factor, the produce harvested from mid- to end-of-season can be "put up" or stored in a variety of ways, from freezing and canning to drying or pickling.

Directions for each method of food preservation are available in many general cookbooks, and in specialty titles available at the library or bookstore.

Freezing is probably the simplest food storage method, as well as the cheapest, since the only store-bought items needed are plastic, zip-lock freezer bags.

Home canning requires glass jars, lids and rings, and a canner— either an inexpensive "hot water bath" metal pot that simply sits on the stove top and boils, or a more pricey pressure canner. My own preference is for the latter, because pressure canning is generally faster and more precise than the hot water bath method. Pressure canners— essentially identical to pressure cookers, but made big enough to hold several quart canning jars—are available from a variety of manufac-turers. If possible, look for a stainless steel model, rather than the more common aluminum canners.

And check to see what sort of weight is used to regulate pressure at the canner's pressure vent. Some manufacturers have a weight that disassembles into three pieces, like an old-fashioned monkey puzzle. One piece set onto the vent will keep the pressure at 10 pounds, two pieces joined together at 15 pounds, and all three together at 20 pounds. The only difficulty is that the pieces are clumsy to take apart, and tend to separate when you try to put them in place or remove them from the vent. Other manufacturers use a simple, round, single-piece weight that has three calibrated holes around the rim, for 10, 15, or 20 pounds pressure. Just set the desired hole over the vent, and there is no worry about the weight falling to bits! Canners all come with clear, well-written instruction books, most of which include a selection of basic recipes.

Most supermarket grocery stores carry canning jars, in cartons of one dozen each, as well as lids and screw-band rings. The Bernardin company sells both an instruction/cookbook and a kit, also available in groceries, that includes a jar lifter, thermometer, and other handy canning gadgets. The same jars one uses for canning are also used in pickling, so whatever method is employed the jars will serve the purpose. And they can be used over and over, year after year. Some of my quart jars have been in use, season after season, for more than 20 years.

The screw rings can also be re-used, although the ordinary sealing lids sold in supermarkets cannot. In the old days, rubber seals came separately from the lids and could be re-used several times. But most lid makers today produce a one-piece metal lid with a thin layer of sealant, usually colored red, around its rim. The sealant won't work if used more than once. Reusable lids are only available via mail order, from a variety of companies that advertise on the Internet. To find them, just type in "reusable lids" and let the Google search engine list them for you. I've used both reusable and one-time-only lids. Both work fine.

Food drying or dehydrating is another easy preservation method, with a selection of moderately priced dryers on the market. The most common are solar food dryers, which require no power source other than sunlight to work, and plug-in electric dehydrators, which are faster, more precise, but also more expensive. I've used both types. The distinction, for dyed-in-the-wool food preservers, is a little like that between crossing the lake in a sailboat or a motor-powered cabin cruiser. Both will get you there, but one is faster and takes less skill to use. There are also plans for build-your-own solar dryers, several of them available free on the Internet. I use dryers to preserve spices like chives or parsley, to dry the herbs for homemade herbal teas, to make dried fruits, and to preserve fish fillets. There aren't many foods that can't be dried, so essentially it's a matter of taste and one's own personal preference.

For more ambitious folks, with a bit more land than average, it's also possible to raise your own grain, and make your own flour. I once had a two-acre garden and kept part of it in vegetables and part in wheat. I bought a mail-order, all-purpose hand grinder (models powered by electric motors are also available) and each fall used a hand scythe to cut my wheat, winnowed it by hand, and then ground the grain into whole-wheat flour, which I stored in airtight plastic buckets purchased at a local food co-op.

It was a real, hands-on taste of how our grandparents' or great-grandparents' generation used to do things, and how people in the underdeveloped countries of the Third World still do this job.

And the homemade loaves baked with that home-grown flour was the best bread I've ever eaten, bar none.

BRANCHING OUT

Individual gardeners, who find their food production options and ambitions limited by lack of space or lack of time, may want to branch out and combine their efforts with those of neighbors. This has been the impetus behind the so-called "City Farming," or Community Gardens movement. The movement began in earnest in North America in the early 1970s, but traces its inspiration further back, to the "Victory Gardens" of the twentieth century's two world wars, when citizens were encouraged to plant home vegetable gardens as a way of dealing with wartime food shortages.

The idea behind it is simple. People who like to garden team up in groups and share both garden space and chores. Where local municipalities are cooperative, they may use city parkland, land along roadsides, vacant lots, or the medians between boulevards or parkways. Green spaces near apartment complexes, the margins of school playgrounds—any place where land is available—may be utilized.

In some cases, gardeners plant individual plots on city land, and have rights to all of the produce from those plots. In other situations, gardeners may share space or work, and afterward divide the harvest among themselves. There are as many possible arrangements as there are people and communities.

In the '70s, the movement was almost exclusively private and grassroots. Today, however, university extension offices, non-governmental organizations (NGOs) and often local municipal governments have gotten involved, and the movement has a thousand faces. The United Nations Food and Agriculture Organization has even entered the picture, encouraging the dispossessed urban poor of the Third World to plant for self-sufficiency in the slums and shantytowns where they've been forced to live, after international trade agreements sandbagged prices for the export farm commodities they used to raise, bankrupted them, and forced them off the land.

Perhaps the easiest way to get in contact with a local community garden group is to phone the Parks and Recreation or Forestry department of your local municipality and ask if they sponsor, or at least know of, any such groups. These days, it's a rare town that doesn't.

In the U.S., for example, cities like Portland, Oregon, have supported community gardening projects for decades. Portland's Parks and Recreation Department (PPR) program has been operating since 1975, and currently has 28 community gardens located throughout the city, operated by volunteers and PPR staff.[10]

In Canada, one of the earliest community garden projects was begun by the City Farmer organization in Vancouver, British Columbia, in 1978, and is still going strong.[11] The program features a nationally famous demonstration garden, which employs the latest in appropriate technology and organic growing methods, from using earthworms to speed composting to building houses out of straw. A recent estimate revealed that as many as 44 percent of Vancouver families grow some or all of their own food, and visitors come from around the world to tour the city's gardens.

Similar projects are thriving in cities across the continent, from Idaho to Ontario.

Some of the better general websites to search for information on community gardening include:

Garden Web,
http://forums.gardenweb.com/forums/commgard
a site posted by Virtual Mirror Inc. of Florence,

Massachusetts, which includes a forum/discussion list where gardeners can seek advice, discuss common issues, and trade horticultural ideas:

City Farmer—Canada's Office of Urban Agriculture
http://www.cityfarmer.org
which has a long list of news items, links, sources of instruction, and a Canadian discussion list/forum on organic and community gardening.

American Community Garden Association
http://www.communitygarden.org,
a national, non-profit association which offers training programs, publications, and information on meetings and conferences across the U.S.

A brief Google search under the search terms "city farming" or "community gardens" will turn up literally hundreds more references and links to local and international urban garden organizations. Chances are good that there is one in your hometown.

NO GREEN THUMB?

If the idea of home gardening isn't appealing (there are some people who just don't have green thumbs, for whom gardening really would represent an impossible challenge, or who are simply too busy to grow their own food), there are other alternatives.

One is to buy your groceries from a community garden group, many of which take on "subscribers" who agree to purchase a certain amount of produce every month, thus providing the group with financial support as well as helping to create stronger neighborhood networks in the community. In some cases, groups will deliver cartons to subscribers' doors, while in others subscribers must come and pick up their consignments at the garden site. Agricultural extensionist Dr.

Joji Muramoto, of the University of California at Santa Cruz, works with a group of community gardeners and educators sponsored by the university's Center for Agroecology and Sustainable Food Systems. Born in Japan, he recalls how his mother helped to set up one of the first food-buying subscriber groups in the Tokyo metropolitan area—an example which sparked his initial interest in organic agriculture and eventually led to his becoming a career agricultural researcher.

"It was in the early 1970s, when all the food safety issues were getting really big, all over the world. When [Masonobu] Fukuoka was writing, also Rachel Carson's *Silent Spring* was translated in Japan. And my mom was very concerned. She was always talking about soils and so on. [laughing] That's the beginning of my 'organic sickness.'

"The group she helped start is still there, buying from the same growers group after 30 or 40 years. There are several thousand households in the Greater Tokyo area that subscribe. The growers deliver door-to-door, produce and deliver. So the families know, face-to-face, who they're buying from."

The university's Community Supported Agriculture (CSA) program isn't on the same scale, and with California's congested highway system, can't afford to deliver. But its members are enthusiastic and bursting with ideas. It is closely integrated with the university's Apprenticeship in Ecological Horticulture program, which trains 40 people each year in sustainable agriculture and organic growing methods. Working under the direction of university staff, the apprentices grow and harvest a wide variety of organic crops, and market them on a "share" basis.

A full share, which feeds four people, costs $600 and provides subscribers with "a diverse supply of freshly harvested organic fruits and vegetables" for 22 weeks throughout the growing season, June to October. That is, a family of four can eat exceptionally well, for roughly $27 per week. Half shares, which feed two, are also available for $380, or about $14 per week. Shares are available at half price for low-income households. Subscribers pick up their consignments weekly at the CSA's barn, where they also have access to a nearby pick-your-own herb and flower garden.

"They [students in the apprenticeship program] are here for six months and they rotate through three different sites, two gardens and the fields, which is tractor-cultivated, about 10 acres of row crops and orchards," says program coordinator Nancy Vail. They currently supply "about 100 households" and "ideally," graduates of the program will be able to return to their home communities and found similar programs there.

Another good source of fresh and, increasingly often, organically grown food, are local farmers' markets.

Until roughly the mid-twentieth century, outdoor farmers' markets were dominant, or at least prominent in most North American cities. In smaller towns, market day was usually Saturday, when farmers from the area immediately surrounding the town would drive in with bushel baskets of things to sell and set them up on tables at the market site. In bigger cities, like Detroit or Chicago or Toronto, one or more outdoor markets were open every day. Detroit's Eastern Market and Western Market, on the city's east and west sides, were the favorite spots for most families to shop, and "truck farming" fresh produce was a reliable source of income for family farms in the region surrounding most cities.

With the coming of the supermarket chains, and the tendency toward vertical and horizontal concentration in the food industry, indoor chain stores with sophisticated packaging, merchandising, and advertising plans, not to mention predatory price wars, gradually took over the industry, forcing out the local farmers and leading to the slow withering of the once-ubiquitous outdoor markets. At one point, it looked like the traditional markets would eventually be wiped out altogether, becoming a mere nostalgic memory from "grandpa's day."

Fortunately for the health of Americans and Canadians, reports of the death of the farmers' markets were, to paraphrase Mark Twain, "greatly exaggerated." In recent years they have seen a revival, spurred in great part by the environmental movement of the 1960s and '70s, which gave impetus to the organic farming movement, and by the general public's growing concerns over personal health.

According to the U.S. Department of Agriculture's Agricultural Marketing Service (AMS), the number of farmers' markets in the U.S. rose by 79 percent from 1994 to 2002, and the numbers are still climbing. A similar development has been taking place in Canada. Many states and provinces have formed farmers' market associations or federations whose aim is to promote and support the growth of this once "traditional" but now cutting-edge marketing vehicle.

In California, for example, there are several associations, some independent and others supported by the university system. Two of the best known are the California Association of Farmers' Markets (www.cafarmersmarkets.com), and the California Federation of Certified Farmers' Markets (www.farmersmarket.ucdavis.edu), the latter sponsored by the University of California at Davis. In Canada, the Farmers' Markets of Ontario has a solid following, and its website features a "market finder" listing local affiliates.[12]

The U.S.-based Local Harvest organization posts a national list of farmers' markets on its web page, as well as organic farms, family farms that sell produce locally, grass-fed meat producers, and other progressive sources of healthy food.[13] The page includes a very useful click-on national map making it easy to find the operations nearest you.

A similar effort is underway to promote traditional, independently owned "country stores" in rural areas, as an alternative to the corporate chain stores and mini malls that have turned so many small towns into cookie-cutter versions of each other. Many of these country stores also feature local and/or organic produce.[14]

As with seed savers and community gardens, hundreds of other farmers' market sources and groups can be located with a simple Google search. Most farmers' markets now include sellers whose eggs come from free-range chickens, and whose dairy products come from grass-fed cows.

Some of the farmers who come to such markets also raise rare poultry or livestock breeds and make them available on order to city dwellers who reserve them in advance. For example, Mary and Rick Pittman of Fresno, California, raise heritage turkey breeds for gourmet cooks who say their birds are tastier and more flavorful than the mass-produced birds sold in supermarkets.[15] So does Sylvia Mavalwalla, of Petaluma. Customers from all over the U.S. order the Pittmans' birds, rather than be content with the mass-produced broadbreasted white turkeys sold in most supermarkets.

Most city dwellers are unaware that, of the thousands of breeds of cattle, pigs, chickens and turkeys that once populated farmsteads around the world, only a pathetic few are still sold commercially by large-scale producers. As Janet Vorwald Dohner explains:

> Today four out of five dairy cows in North America and Europe are the familiar black-and-white Holstein-Friesians. Many people probably believe that all dairy cows are black and white. Moreover, owing to the wonders of artificial insemination, many Holstein-Friesians share the same bloodlines. It may seem hard to believe that just 80 years ago there were 300 breeds of cattle in Europe and North America.
>
> The farmyard pig has today been transformed into a grain-fed, fast-growing meat producer. Just three breeds or their crossbred offspring supply the majority of the market. They are increasingly raised in huge indoor complexes on contract to one of a handful of processors.
>
> Almost all eggs and poultry meat today come from a few hybrid chicken strains [mostly white leghorn], and essentially one type of turkey is raised commercially. Owing to the enormous development of desirable breast meat, turkeys are no longer able even to mate naturally. A limited number of company-owned strains of these special-

ized hybrids supply 95 percent of the North American and European
poultry food source, whereas 50 years ago thousands of individual
hatcheries in North America each raised many breeds of poultry.[16]

With typical shortsightedness, corporate/industrial food producers concentrate their efforts on one or two breeds which appear—at least on the surface—to be the fastest or highest-volume producers. All other breeds, regardless of their particular virtues in terms of flavor, resistance to disease, or adaptation to local or regional environmental conditions, are simply discarded—or, more accurately, wasted. Accountants, not farmers, make the decisions.

Shoppers who patronize their local farmers' markets thus not only assure themselves of a more healthy supply of food for their own families, but also go a long way toward supporting the survival of the continent's embattled independent family farmers, and of the genetic diversity of both the plants and animals who share our planet.

Many family farmers have been quick to catch on to the potential of farmers' markets and community buyers' groups. The once endangered profession of truck farmer is enjoying a small renaissance as these now mostly organic truck farmers multiply and return to the marketing fray.

Of course, this provides an obvious opportunity for the more ambitious among those concerned with food quality, those intrepid city folks willing to follow the lead of the now-legendary Helen and Scott Nearing, who went "back to the land," and wrote about it in their now-classic best-seller, *Living the Good Life.*[17] The Nearings' work inspired an entire generation of back-to-the-landers in the late 1960s and early 70s, spawning national magazines like the U.S.'s *Mother Earth News*, Blair and Ketchum's *Country Journal*, and Canada's *Harrowsmith*. Many of those who emulated the Nearings are still in the country, still living the good life, and wouldn't mind a few new neighbors moving in next door.

Although not a vegetarian like the Nearings, the author of this book plans in about a year to rejoin the movement of which he was once a part (during the '70s), returning to the farm life he enjoyed in Quebec and Ontario.

But that's another story, for another book.

Think Locally, Fight Locally   8

BY ITS VERY NATURE, INTERNATIONAL CORPORATE agribusiness cannot be "reformed," or pressured into becoming a reliable, responsible source of healthy food and a protector of the environment, any more than foxes could be trained to become guardians for the world's chicken coops. Corporations, especially American corporations, have only one concern: maximizing short-term profit, usually on a quarterly basis, along with the salaries and other perks of their upper management teams. Anyone who doubts that need only look at the recent example of Enron, or any number of other glaring embodiments of corporate amorality.

In Western Europe and Japan, corporate executives tend to look a little further beyond the next business quarter than do their American counterparts, and their personal compensation packages tend to be a bit less outrageous, but even there, any corner will be cut to maximize profits. Following the rigid, neo-liberal ideology of the Chicago School's economists, corporations have one goal only: profit. Everything else, including your health and mine, despite all the feel-good advertising about "feeding the world," is just an annoying distraction.

And the largest corporations are effectively unaccountable for their policies, for two reasons:

1) although they are in fact no more than legal agreements printed on pieces of paper, meant to shield their executives and shareholders from liability for illegal or other damaging acts they might commit, American corporations have since 1886 been regarded under the law as the equivalent of human "persons," with all of the rights to which living, breathing people have a claim. Apologists for this system like to assert that this was a result of a U.S. Supreme Court decision, in the case of Santa Clara County vs. The Union Pacific Railroad. In fact, it wasn't. The court itself made no such judgement. However,

> The Recorder of the court—a man named J. C. Bancroft Davis, himself formerly the president of a small railroad—wrote into his personal commentary of the case (known as a headnote) that the Chief Justice had said that all the Justices agreed that corporations are persons. And in so doing he—not the Supreme Court but its clerical recorder—inserted a statement that would change history and give corporations enormous powers that were not granted by Congress, not granted by the voters, and not even granted by the Supreme Court. Davis's headnote, which had no legal standing, was taken as precedent by generations of jurists (including the Supreme Court) who followed and apparently read the headnote but not the decision.[1]

The result has been to give corporations enormous legal powers, in addition to their already massive economic clout, elevating these abstract entities' rights to the point where they actually supersede those of real people;

2) as anyone who has followed politics in recent years, and who is not hopelessly naive, knows only too well, national government in the U.S., and to a slightly lesser degree in Canada, is government of, by, and for a cabal of rich, mostly white men who have been or still are CEOs of huge, transnational corporations, or who have represented

these corporations as their lawyers. No one other than a millionaire, or someone supported by and indebted to millionaires, can afford the huge monetary outlays needed to buy media time and space and run for national office.

Repeated exposés, by any number of writers in any number of venues, have revealed the ties that bind and blind.[2] Presidents, senators, congressmen and women, prime ministers and members of parliament (a few members of Canada's New Democratic party may be exceptions) are, for the most part, so many political whores, on the indirect payroll of the special interests who financed their election campaigns, and who will finance their lucrative post-political careers via well-paid consultancies, book publishing deals, and speaking fees.

The combination of virtually unaccountable economic, legal, and political power amassed by corporate North America has almost totally eclipsed that of struggling family farmers and ordinary consumers. It is not the goal of this book to detail the precise historical hows and whys of this tragedy, which have been reported elsewhere by investigators like the American researcher A. V. Krebs and the Canadian Brewster Kneen, whose ongoing work is a tribute to unselfish, dedicated concern for the public good.[3] But the results of this socio-political development are undeniable.

At national level, ordinary people who want to protect themselves and their families have almost no voice, because those who are nominally supposed to represent them are already "owned and operated." At best, these political whores will give lip service to a popular cause during an election, but once elected they'll drop it like a hot potato and toe the corporate line.

Meanwhile agribusiness and its related industries go their way, inadequately regulated. A recent example of how weak federal regulation in general is, was provided by the *Journal of the American Medical Association*, which published a strongly worded editorial attack on the U.S. Food and Drug Administration's record as a watchdog. One of the co-signers of the December 2004 editorial, Dr. Drummond

Rennie, later told an interviewer: "There is no question that the sen-
ior folk at the FDA are behaving as if their primary client was the drug
companies and not the public."[4]

Corporate America chases one get-rich-quick scheme after another. When a problem appears with one of these schemes, the tendency is not to take the time to investigate it thoroughly, but to look for a
"get-rich-quick fix," and the simpler that fix is, the better. Too often,
fixing problems this way only means creating new problems.

Much of the push to introduce genetically modified crops and
livestock has stemmed from this approach. The desire not to "waste"
time or money rotating crops from one season to another, as a means
of controlling weeds or insect pests, helped spur the development of
chemical pesticides and herbicides. The idea was to maximize production of a given highly profitable crop, year-after-year in the same
fields, using chemicals as the panacea for any pest outbreaks that
resulted. When insects and weeds developed resistance to these chemicals, still more lethal compounds were introduced, until the compounds threatened to endanger the survival of the crops themselves.

To the rescue came the gene manipulators, who created crop varieties that would not be affected by the toxic chemicals, or which themselves were toxic to the pests in question. Unfortunately, the new varieties could not be contained in a given industrial farm field. Their seed
and pollen tended to blow on the wind, onto neighboring fields,
sprouting where they were not wanted or cross-pollinating with other
varieties. Virtually indestructible due to their herbicide resistance,
they became, in effect, superweeds that neighbors could not eradicate,
contaminating farm after farm and endangering the purity or even the
survival of other plant strains.

One of the most frightening examples of this phenomenon was
the discovery of "escaped" GM corn (maize) in Mexico. Mexico is the
geographic center of origin of maize, the place where the plant first
evolved and where it is crucial to maintain pure original strains as a
genetic "bank" for future use. In 2001, GM varieties of maize were dis-

covered in two Mexican states, creating an international outcry. Subsequent genetic tests of maize grown by farmers in 138 Mexican communities showed contamination actually

> ...has spread to farmers' fields in nine states, including Chihuahua, Morelos, Durango, Estado de Mexico, Puebla, Oaxaca, San Luis Potosi, Tlaxcala, and Veracruz.
>
> Of 2,000 maize plants tested, samples from 33 communities in nine Mexican states tested positive for contamination. In some cases, as many as four GM traits, all patented by multinational gene giants, were found in a single plant. The organizations were especially alarmed to find traces of the insecticidal toxin (Cry9c), the engineered trait found in StarLink maize (formerly sold by Aventis CropScience). StarLink was never approved by the U.S. government for human consumption because of concerns it could trigger allergic reactions. Illegal traces of StarLink were found in U.S. food products in 2000. Following a massive recall of tainted food products in the U.S., Aventis withdrew StarLink from the market.[5]

Similar cases of contamination by GM crops have been reported in Canada, including instances where GM wheat, soya, or canola (rapeseed) have spread to neighboring fields, endangering the livelihoods of local farmers.

Meanwhile, continued use of the toxic chemicals GM varieties were engineered to tolerate took an increasingly heavy toll not only on plant pests, but on beneficial insects and plants as well. Tests in Britain in 2003, for example, found that "two of three types of GM crops tested in farm scale trials are worse for wildlife than conventional crops."[6] The evaluations, which were "the largest of their type ever carried out," found that "GM beet and spring oilseed rape reduced the numbers of some birds and insects as the field trials hosted fewer weeds and weed seeds that support wildlife." Noted research team head Dr. Les Firbank: "If the trends we have seen continue, then we could see long-term declines of these weeds that are important food sources for birds."

Plants genetically manipulated to be toxic to certain insect pests also present a potential threat to non-pest species, such as honeybees, butterflies, or insects crucial as a source of food for birds or other wildlife, or needed for pollinating major farm crops.

Eva Novotny, of Scientists for Global Responsibility, summed the situation up succinctly in an article in Britain's *Guardian* newspaper:

> *Science has reached a point where the imagination and technical capabilities of scientists are overtaking society's ability to evaluate and control the outcome. The perception of many scientists is that all that can be done in science should be done—and if we do not do it, a competitor will. But their theoretical models of the natural world do not encompass the complexities of the real natural world. Nature works in profoundly subtle, intricately balanced, and interconnected ways that we do not yet fully understand. That is why independent scientists urge caution before we release into the environment, and into our bodies, crops and foods that have been developed by crossing not only dissimilar species but even kingdoms. The long-term consequences cannot be predicted.[7]*

The quick fixes and get-rich-quick schemes of corporate North America, by their very internal logic, which is the logic of pure greed, are incapable of assuring the best interests of the public, that is, of you and me. And our national governments are unlikely to exert much of a brake on their headlong rush for imagined bonanzas.

GRASSROOTS POWER

So where do those who are justly frightened by the current situation in the continent's food industry go to start turning things around, and to confront corporate power?

The answer, at least in the short run, is not to confront it but to go around it. A basic principle in the martial art of aikido is to avoid head-on clashes, to move obliquely, neutralizing the opponent's strength by marginalizing it. The way to defeat corporate power is to make it as irrelevant as possible to our daily lives by setting up our own, alternative systems and insulating those systems effectively while they grow and mature.

Look at the Amish people: They've been doing it for decades. Granted, they've gone a bit farther than most of us would want to in rejecting modern social mores and technology. Beards and horse-drawn buggies aren't exactly everybody's style. But the Amish and Mennonites have certainly managed to construct an alternative, self-sufficient reality that allows them to live as they choose, relatively free of dependence on anyone else's system.

Without rejecting truly appropriate technology, or trying to turn back the clock to the nineteenth century, others could do the same. Home gardens and community gardens planted with heritage seeds are a first step. Patronizing farmers' markets, independent country stores, and supporting local truck farmers are the next steps toward creating an alternative food system, an alternate reality.

The place to start, as with most long-term projects, is in our own neighborhoods and local municipalities. To paraphrase the famous 1960s adage "think globally, act locally," the immediate goal should be "think locally, act locally."

Direct political and economic confrontation are also possible, at the local level, long before there is enough strength to go national. The cost of a political campaign for local offices, such as city council, county commission, school board, or other, similar positions, is not so great as to exclude anyone but millionaires or those supported by the sinfully rich. Ordinary people, backed by relatively small groups of citizens, can still afford the price of entry to this sort of political contest. And once installed, they and those of like mind can have a real impact.

Municipal policies are effective tools, as well as potent statements of public opinion. Take the matter of gene-altered or genetically manipulated foods, whose rapid introduction national governments seem to have adopted as an unquestioned priority. In California—the very jewel in the crown of corporate agriculture—voters in Marin County, just north of San Francisco, approved a November 2004 ballot proposition banning the growth or raising of genetically engineered plants and animals within county borders.[8] Marin thus became the third county in the state to ban GM crops, joining Mendocino and Trinity Counties, which enacted bans earlier in the year. According to *USA Today* reporter Elizabeth Weise:

> *The measures mark what's likely to be a growing grassroots fight over the use of bioengineered crops in the state and nationwide. Signature gathering has already begun for an anti-biotech crop measure in Sonoma County, a heavily agricultural county in California's wine country.*
>
> *Renata Bellinger, director of Californians for GE-Free Agriculture, says she knows of "at least a dozen" other counties where activists are considering launching campaigns.[9]*

Make no mistake, such grassroots campaigns give an example to other communities, and as they spread can eventually determine the entire direction of government policy. In the European Union, GM foods, with very few exceptions, are barred—not because federal higher-ups with compromising ties to industry want it that way, but because ordinary Europeans simply will not stand for the introduction of the crops. Countries like France and Italy have centuries-old traditions that govern their cuisine and views of food quality. Even if government should permit GM foods in, shoppers simply would not buy them.

Chances are a lot of North Americans would not buy these foods either, if they knew which products on the grocery shelves contained them. But they don't. The fact that manufacturers have not been required to label GM foods has enabled corporations to introduce a host of GM-laced products to North American supermarket shelves without consumers' knowledge. As the *International Herald Tribune* reported:

> The majority of corn and soy in the United States is now grown from genetically modified seeds, altered to increase their resistance to pests or reduce their need for water, for example. In the past decade, Americans have happily—if unknowingly—gobbled down hundreds of millions of servings of genetically modified foods. The U.S. Food and Drug Administration says there have been no adverse effects, and there is no specific labeling.[10]

That, by the way, is the same FDA referred to by the authors of the *Journal of the American Medical Association* editorial, mentioned previously.

Campaigns have been launched to force grocers and food manufacturers to clearly label food products that contain bioengineered components. The movements behind these campaigns are basically grassroots in nature, or driven by non-profit, non-governmental organizations like the Center for Science in the Public Interest. They are growing fast.

Consumer boycotts have also proven to be a potent weapon in the battle against overweening corporate power. The grape boycott of the 1960s, alluded to in Chapter Four, had a tremendous political impact, eventually forcing growers to recognize Cesar Chavez's United Farm Workers union.

As the Manchester *Guardian* reported in a recent article, "consumer boycotts cost big brands $4.92 billion a year," and some sharply focused boycotts have had quick results.[11] As the article noted: "The X factor in last year's Burma Campaign boycott of Triumph, the lingerie manufacturer, was the advertising. It showed a model wearing a barbed-wire bra under the slogan 'support breasts not dictators.' Customer complaints and reports of middle class women returning their bras to Selfridges proved too much for Triumph—it was out of Burma within two months."

If federal-level politicians are at the moment unresponsive to the wishes of ordinary people, because the major corporations that finance their election campaigns are telling them to ignore the issues, that does not mean that we cannot have a local impact. And as the number of local impacts grows, making it more and more difficult for national policies to take effect on the ground, state and provincial governments will begin to take notice as well.

Of course, sometimes the notice taken is not the kind grassroots organizers want to attract. Fearing public exposure and wanting to head off criticism of factory farming practices, the food industry has come up with a variety of legal tactics to harass its critics. Most recently, these have included a new type of legislation, which may be the most brazen attempt in American history to stifle free speech, namely so-called "food disparagement" laws. As the authors of *Fatal Harvest: The Tragedy of Industrial Agriculture*, note:

> The industry has pressured 13 states to pass "food disparagement" legislation—laws that can be used against those trying to expose any of the harmful effects of the industrial food system. While many believe these laws are clearly unconstitutional, until they are struck down, they serve to intimidate people and groups who want to provide truthful information on food safety.[12]

These laws, in effect, make it illegal to speak, write, or broadcast much more than steady praise about food or food products. In theory, at least, a restaurant critic could be sued in some U.S. states for giving an honest review of a local eating establishment. Oprah Winfrey was even a victim of these laws, when a group of meat industry executives sued her and a guest who had made adverse comments about hamburger on her national television show.

When this news reached Canada, one wag in a newspaper office called over to the paper's theater critic: "Hey, if you pan an actor, do you think they'd sue you for ham disparagement?" Everyone laughed, but the situation is closer to reality than most of us would like to admit.

However as the saying goes, "truth will out," and when not only municipal but state and provincial legislators begin to weigh in on the side of the people, federal politicians will have to listen whether they want to or not. Especially in Canada, where the rules of Confederation give the provinces far greater powers vis-a-vis the federal government than U.S. states have over Washington, provincial premiers form a bloc that even the Prime Minister of a majority government cannot ignore. A minority government is still more vulnerable.

But the work of bringing change at the top has to start on a firm foundation, and be built from the ground up.

TUNE IN, FIND OUT

A lot has been said about the educational and networking power of the Internet. Perhaps a bit too much has been said, to the point where it's easy to tune it out as mere hype. But it's not hype. Anyone who wants to learn what is really going on in the food industry need only look as far as their PC. A host of non-governmental organizations, citizens groups, consumer organizations, and family farm associations are out there, watching the every move of the food industry's corporate players, ready to pounce, fighting the good fight, and striving for the good of ordinary people and their families. Some, like the National Farmers Union (NFU) in the U.S. and Canada, track pending farm-and-food related legislation, alerting visitors to their websites and helping rally support or opposition, as each case merits.

They are only as far away as a Google search, or the nearest Internet bulletin board or discussion group. If you can find eBay, you can find them.

In Chapter Ten, a list of sources—of tools, practical information, guidance, networking possibilities and groups working for the common nutritional good—is provided. It is far from comprehensive, since the Internet and the world of progressive activism changes daily. But it ought to provide at least a starting point.

Before looking at that list, however, it may be worthwhile to reflect on what this whole controversy over food is really all about.

THERE USED TO BE A FARM IN HARROWSMITH,
Ontario, where we lived and had friends, as well as a big, round, solid
oak table in the kitchen of our house. One Christmas Eve, we sat
down at that table for dinner and the menu consisted of ruffed grouse,
or partridge as most people in Frontenac County refer to them, shot
on the wing in the cedars we were raising in our woodlot to sell to a
local lumber yard as fencewood. Maybe it would have been more traditional to have put those birds up from a clump of pear trees, but the
cedars had done just as well. The recipe, taken by the cook from the
Canadian Wild Game Cookbook, then adjusted to fit her own preferences, was too complicated for me to remember (being probably the
worst cook in the world, I have trouble remembering how to heat
canned soup). But they were delicious.

On the same plate with the partridge were carrots and green pole
beans, raised in our own one-acre garden and harvested that fall. We'd
had to struggle a bit to protect the garden from nibbling rabbits,
groundhogs, and a wide assortment of crawling, skittering, and
chomping things, from cabbage worms to slugs, but in the end we had
enough vegetables frozen, canned, or dried, along with dried herbs

with which to flavor them, to last the winter. The kids and I had spent many hours between the rows, weeding, cultivating, mulching, the kids grumbling. And after the last harvest in the fall, I'd plowed the garden down with our old, all-purpose International diesel tractor, incorporating the non-edible residue to nourish the soil for the following year. Not one drop of chemical pesticide or herbicide had touched the garden since we bought the land.

On the same plate with the partridge, carrots and beans, was a generous helping of *Zizania aquatica*, a.k.a. wild rice, harvested that year on Mud Lake by my friend Band Chief Harold Perry, in the traditional manner of North America's native peoples. He'd gone out in his canoe, threading his way through the rice stands, bending the stalks of grain over the edge of his canoe and thrashing the grains off of them with two cedar sticks, letting them fall into the canoe. We'd hunted ducks together that season, Harold, my son Ed, and I, and he'd given me a big, bulging sack of the precious rice, worth a fortune to gourmet diners at retail, for nothing. The Ojibway people call it manomin, the delicacy of the Great Spirit. They're right.

Also on the table was freshly baked whole wheat bread, made from flour I'd gound myself in a hand grist mill, from wheat I'd scythed by hand from my own acreage and later winnowed over an old bedsheet. To drink after the meal, we had white clover tea, made from blossoms picked from the yard and then dried, as well as vitamin-rich dandelion coffee, made from the oven-baked roots of the ubiquitous yellow-flowered "weed." The milk came from two cows, an Ayershire and a Jersey, owned by our neighbors, Dorothy and Claud, down the road. It was probably a bit too rich in butterfat, thanks to the Jersey.

Although in the following year I made my own sour-mash corn whiskey (illegal, of course) this year's after-dinner drink was French cognac. Along with some bakery fruitcake, it was the only store-bought thing in that entire meal.

More than 20 years later, I still remember that dinner, not so much for what was in it, but for how it got there. We'd put everything there ourselves, by our own skill and labor. Food acquired that way has an added flavor all its own, to accent any holiday air.

WITH DUE ATTENTION

But food need not be home-grown, or home-shot, or harvested with a canoe to be worth eating. I recall with almost equal pleasure a meal shared one summer evening with my niece Annette and her husband Martin at Chez Panisse, Alice Waters's famed restaurant in Berkeley, California. The restaurant is named after a character in French playwright Marcel Pagnol's classic *Fanny*. Ms. Waters's career as a gourmet restaurateuse was inspired by her early experiences with French cuisine, and her love of French culture is reflected in the name of her establishment.

Being a bilingual Canadian, who as a young man spent a year in France, where I saw the original 1930s movie version of *Fanny* with the legendary actor Raimu playing the part of Fanny's father, Cesar, I found the place congenial. It got even more congenial when the waiter turned out to be a young Frenchman, from the same Midi where Pagnol once lived. I ordered in French, with the waiter chuckling at my Quebecois accent and me chuckling at his half-Italian Midi drawl.

The food, of course, was superb. I had blanquette de veau— "creamy braise of Mavalwalla Ranch pasture-raised veal with morel mushrooms and Chino Ranch carrots." Note the name Mavalwalla, the same California farm mentioned in the previous chapter where one could obtain rare livestock breeds. The green salad I had before the meal was the crispest and tastiest I've ever had in a restaurant. Both of these points are in keeping with Ms. Waters' cooking philosophy, namely to serve "only the highest quality products, only when they are in season." As her website explains:

> Over the course of three decades, Chez Panisse has developed a network of mostly local farmers and ranchers whose dedication to sus-

*tainable agriculture assures Chez Panisse a steady supply of pure
and fresh ingredients.*

*Alice is a strong advocate for farmers' markets and for sound and
sustainable agriculture. In 1996, in celebration of the restaurant's
twenty-fifth anniversary, she created the Chez Panisse Foundation to
help underwrite cultural and educational programs such as the one
at the Edible Schoolyard that demonstrate the transformative power
of growing, cooking, and sharing food.*[1]

The key word there is "sharing." Food is not just something you
jam into your mouth and swallow fast to prevent starvation. It is the
basis of social interaction. From a baby's first bonding with its own
mother, through the milk from her breasts, human beings have used
food as a means of keeping family, clan, and community together. In
every religion and every culture, around the world, sharing a meal has
been seen as a necessary social component of important occasions:
whether at Christmas, Easter or Iftar, at weddings, baptisms, or wakes,
sharing a meal is the key to sharing life.

Pressed by the demands of work and daily cares, we may not
always be able to give this ritual its due attention. But it should be
given much more regard than it is in our present culture. To make the
neglect of food a habit, its production a mere conveyor-belt, assembly
line routine measured in some corporate ledger book, and the eating
of it a peripheral event to be gotten through quickly, is to make it a
habit to forget what makes us human.

Our species is social. As Lynn Margulis indicated in her pioneering
books, *Origin of Eukaryotic Cells*, and *Symbiotic Planet*,[2] it may be precisely this social, cooperative, communal attribute which has guaran-
teed our species' survival—in truth, the survival of all biological life.
We survive through cooperation. Our cooperation is symbolized and
accented each time we share a meal with another human being, and
consciously take the time to enjoy it together.

Whether the feast is Jewish, Muslim, Hindu, Christian, or secular, a gathering of family or of friends or of a community, it should be prepared with care, appreciated carefully, and remembered long afterward. My son Ed and his friends, God bless them, know this. When he makes his Polish-style *bigos*, or hunter's stew, and one of his buddies brings his own homemade wine to the meal, they are renewing more than their appetites.

This is the thinking behind the now widely known Slow Food Movement. Launched in Italy, home of the world's best pasta, tomatoes, and (the French will never admit this) red wines, the idea behind it was to be the opposite of fast food. As Michael Pollan writes in *Mother Jones* magazine:

> *[Slow Food] took shape 17 years ago in the brain of Carlo Petrini, a left-wing Italian journalist dismayed by the opening of a McDonald's on the Piazza di Spagna in Rome—and perhaps equally dismayed by the hangdog dourness of his comrades on the left. After years of activism he had come to the conclusion that "those who suffer for others do more damage to humanity than those who enjoy themselves. Pleasure is a way of being at one with yourself and others." So rather than picket McDonald's new outpost in the heart of Rome, or drive a tractor through it a la [French family farm advocate] Jose Bove, Petrini organized a group of like-minded activists-cum-sybarites to simply celebrate all those qualities that McDonald's inexorably drives toward the homogenization of world taste: the staunchly local, the irreplaceably unique, the leisurely and communal.*
>
> *Seventeen years later, McDonald's is still serving Happy Meals by the Spanish Steps (though Petrini did persuade the company to hold the golden arches), yet Slow Food has emerged as a thriving international organization with more than 65,000 members in 45 countries.*[3]

It should come as no surprise, perhaps, that Alice Waters is one of them.

Petrini and Waters have got it right. Food is international, universal, but at the same time ought to be intensely local and individual, like the human beings who produce it. Charles de Gaulle once famously complained, in mock exasperation: "How can anyone possibly hope to govern a nation with a thousand cheeses!" But the same General de Gaulle also said, of that same, hopelessly individual, contentious France, *"Ah mere, telle que nous sommes, nous voici pour vous servir!"* (Oh mother, such as we are, we are here to serve you!) And he risked his life, many times, for his beloved France.

Food is not meant to be hoarded, but shared. No one is more zealous in spreading abroad the goodness of the French table than the French themselves, or the Chinese or the Italians or the Lebanese. Each national, regional, or local cuisine is a treasure, to be protected and strengthened, but not isolated. To truly thrive, it must be shared.

This includes the cuisine of North America's indigenous peoples, from the manomin of the Ojibway to the bannock of the plains Cree. When Euell Gibbons published his classic *Stalking the Wild Asparagus* in 1962,[4] introducing Americans to the culinary joys of indigenous plant species, he caused a sensation. No one in the then-dominant white, basically European culture had thought much about native plants as sources of food. With the exception of a few staples they'd appropriated from the "Indians," like corn and squash, only the familiar, European garden vegetables their ancestors imported with them when they came to the New World were truly perceived by most people as real food.

But the native peoples of the Americas had been here for thousands of years before the white interlopers came, and had had plenty of time to sample, experiment with, and improve upon the fruits and vegetables that had been growing here for eons. In each of the major North American climate regions, entire ecosystems had evolved, and native people understood them, literally, from the ground up. The grasses that fed the plains bison, the main food staple of the Cree and Sioux, were known and understood, along with the herbs, fruits, and flowers of the Great Plains. In the east, the Iroquois and Mohawk had developed deliciously varied cuisines of their own, using the plants they raised in their own gardens, as well as those they gathered in the wild.

This cuisine still exists, and the majority population of North America is the poorer for not knowing about it. I've eaten bison, and bannock, and manomin, and am the richer. In most cases, the foods utilized by the "Indians" were and are also utilized by wildlife, and the gardener who fosters them fosters bird and animal life along with them.

It's all part of what many peoples on many continents have called "the Great Circle" of life. The earth holds the potential for an almost infinite diversity, at the same time as the potential for a great unity. The two are not mutually exclusive, but complement each other.

We don't need to turn cows into cannibals, or lace our meals with tiny, horrid molecular machines, or destroy the soil, in order to create a higher quarterly return for some greedy set of imbeciles in hand-tailored suits and Gucci ties, sitting smugly around the boardroom table in a highrise somewhere. We don't need to despoil the earth and end up eating soylent green.

We need to take back control of our own food supply, our own meals, and our own humanity.

When the first *Whole Earth Catalog* was published in 1968, it came as a revelation to a generation of idealistic young people. They were looking for ways to change the society around them, and more often than not, ran down blind alleys and false paths that led them nowhere. But here was something solid, something practical, something a few of them, who were serving in the then-new and untried Kennedy-inspired Peace Corps, found had answers even to their pressing problems in the field. It also served as a key source for the Back-to-the-Land movement, and for the movement's first cousins, the City Farmers, in the 1970s.

The catalog was a runaway success, and continued to be popular for years, until the idealism of the '60s fizzled out in the Reagan years, the hippies got married, had kids, and their younger sisters and brothers became something called "Yuppies"—or young, upwardly mobile professionals. Investment portfolios became more important than weeding, and an entire society made a sharp right turn.

But the old idealism of the '60s didn't really die. It only went into hibernation. It's time it woke up again, into a new spring. The *Whole Earth Catalog* is no longer with us, having gone out of print following its last annual edition in 1996. But many of the sources cited in it, along with brand new ones, are still out there, available. Canada's *Harrowsmith* magazine, which took its inspiration from such sources, is also no more, but its spirit lives on in places like Frontenac County, where I lived while working as a *Harrowsmith* editor and where I plan to return.

For those who would like to renew the genuine impulse for good that those years represented, and who would like to give themselves and their families a more healthy, truly human life, the following sources are noted. Some of the books may be out of print now, but most are still available at public or university libraries, or online through such organizations as Abebooks or Alibris, which specialize in out-of-print titles. The list is far from exhaustive. It is only my own, eclectic, personal selection.

GARDENING INFORMATION

Fern Marshall Bradley and Ellis, Barbara W., eds. *Rodale's All-New Encyclopedia of Organic Gardening: The Indispensable Resource for Every Gardener*. Emmaus, Pennsylvania: Rodale Books Inc., 2004.

That "indispensable" in the title is not hype. This really is the best book in existence on the subject, for anyone from beginning backyard gardener to professional truck farmer. I've used it since 1973, and can't say enough good about it. For that matter, nearly everything the Rodale organization publishes, including *Organic Gardening* magazine, is first-rate. Visit their website at (www.rodale.com) or contact them at: Rodale Inc., 33 East Minor St., Emmaus, Pennsylvania 18098 USA. (610) 967-5171.

Tanya L. K. Denckla. *The Gardener's A-Z Guide to Growing Organic Food*. North Adams, MA: Storey, 2003. A good general guide, useful for its garden "troubleshooting" section.

Stu Campbell. *Let it Rot! The Gardener's Guide to Composting*. North Adams, MA: Storey, 1998. Some of this book's information is also found in the Rodale encyclopedia, but some isn't. A thorough exploration of garbage and other useful material.

Carole Rubin. *How to Get Your Lawn & Garden Off Drugs.* Madeira Park, B.C.: Harbour Publishing, 1990. Read this. Get the monkey off your back.

Bill Merilees. *Gardening for Wildlife: A Guide for Nature Lovers.* Vancouver, B.C.: Whitecap Books, 2000. How to attract not only birds, but butterflies, frogs, squirrels and raccoons to your backyard. That is, assuming that you don't mind some of the attractees munching on what you wanted for yourself!

Trevor Cole, ed. *What Grows Where in Canadian Gardens.* Toronto: Dorling Kindersley, 2004. A very handy guide to deciding where best to site various garden species, in shade, full sunlight, clay or sandy soil, next to the watercress or not. Though written for Canadian conditions, much of the information is transferable to the U.S.

Lorraine Johnson. *The Ontario Naturalized Garden: The Complete Guide to Using Native Plants.* Vancouver, B.C.: Whitecap Books, 1995. A good, detailed guide to making best use of plants native to North America, rather than those imported from Europe.

J. T. Garrett. *The Cherokee Herbal.* Rochester, Vermont: Bear & Company, 2003. An introduction to the vast store of neglected knowledge of this continent's native peoples concerning the green world around us.

Suzanne Ashworth. *Seed to Seed: Seed Saving and Growing Techniques for Vegetable Gardeners.* Decorah, Iowa: Seed Savers Exchange, 2002. The best book available on the art of saving and replanting heritage seeds, from year to year. By "the best," I mean the best.

Jennifer Bennett. *The Harrowsmith Northern Gardener*. Camden East, Ontario: Camden House, 1982. Techniques for raising food greens where the frost comes sooner and leaves later.

FOR MORE SERIOUS GROWERS

Those who are interested in moving beyond the backyard garden and getting into either professional truck farming or what once used to be termed "homesteading," the following titles may be helpful:

Richard W. Langer. *Grow It! The Beginner's Complete In-Harmony-with-Nature Small Farm Guide*. New York: Avon Books, 1972. For back-to-the-landers of the 1970s, this was a lifesaver. Today, it is also a kind of artifact, giving a glimpse into the cultural mindset of the so-called "movement years," and what motivated all those idealistic youngsters who are now going grey.

Frank D. Gardner. *Traditional American Farming Techniques*. Guilford, Connecticut: The Globe Pequot Press, 2001. A reprint of a text first published in 1916, when there were a lot more family farmers than there are today, this book contains a wealth of nearly-lost practical information on how to raise and process crops as it was done before the age of industrial agriculture. Some of the science is outdated, but most of the advice in this book is still bang-on for anyone interested in organic growing methods. Our grandparents were no fools. Using these methods, they survived and prospered–otherwise, we wouldn't be here.

Lincoln C. Pierce. *Vegetables: Characteristics, Production and Marketing*. New York: John Wiley, 1987. An excellent, professional introduction to vegetable crop production, written by a professor at the University of New Hampshire.

V.I. Shattuck and M. McKnight (revised by A. McKeown and M. McDonald). *Vegetable Crop Culture.* Guelph, Ontario: University of

Guelph Department of Plant Agriculture, 2003. Referred to modestly as mere "course notes" for students in Horticulture 3510 at Guelph University, this is one of the best horticultural texts extant, written by professors at one of Canada's premier ag schools.

Tracey Baute, ed. *Agronomy Guide for Field Crops.* Toronto: Ontario Ministry of Agriculture, Food and Rural Affairs, 2002. Covering everything from corn and soy to canola, this is a reliable text for serious small farmers. Particularly valuable is its section on crop rotation.

Roger B. Yepsen, Jr. *Trees for the Yard, Orchard and Woodlot.* Emmaus, Pennsylvania: Rodale Press Inc., 1976. As a one-time tree farmer (cedars for fencing and maple sugarbush), I found this title very useful.

Claudia Weisburd. *Raising Your Own Livestock.* Englewood Cliffs, NJ: Prentice Hall, 1980. How to raise cows, goats, swine, sheep, chickens and horses—on a small scale.

Patricia Cleveland-Peck. *Your Own Dairy Cow: Essential Guidelines for the Management of a House Cow.* Wellingborough, Northhamptonshire, UK: Thorsons Publishers, 1979. A Brit weighs in on how to maintain the family cow.

F. Lebas, P. Coudert, R. Rouvier, and H. De Rochambeau. *The Rabbit: Husbandry, Health and Production.* Rome: Food and Agriculture Organization of the United Nations, 1986. Like rabbit stew? Here's how to raise it.

Dadant & Sons, eds. *The Hive and the Honeybee.* Chicago: Dadant & Sons, Inc., 1975. Apiculture—the raising of bees for honey and crop pollination—covered in depth by the editors and contributors of the *American Bee Journal.*

SUSTAINABLE AGRICULTURE, IN GENERAL

Masanobu Fukuoka. *The One-Straw Revolution*. New York: Bantam Books, 1985. The best-known title by Japan's world-renowned exponent of organic and traditional farming, this best-seller influenced a generation of growers not only in Asia but around the globe.

Masanobu Fukuoka. *The Road Back to Nature: Regaining the Paradise Lost*. Tokyo: Japan Publications, Inc., 1987. Fukuoka compares Japanese traditional methods with those of California agribusiness. Guess who comes out second-best.

Masanobu Fukuoka. *The Natural Way of Farming*. Tokyo: Japan Publications, Inc., 1985. The author follows up, in greater detail, on the principles first espoused in *The One-Straw Revolution*. Do I like Fukuoka? Whatever gave you that idea....

Stephen R. Gliessman. *Agroecology: Ecological Processes in Sustainable Agriculture*. Boca Raton, Fla.: Lewis Publishers, 2000. Written by the founder of the University of California at Santa Cruz's Agroecology Program, this text introduces the basic principles behind farming as if the farmer was part of the ecosystem, not the system's enemy.

Board on Agriculture, National Research council. *Alternative Agriculture*. Washington, D.C.: National Academy Press, 1989. The U.S. government rediscovers–and endorses–traditional organic farming, roughly two decades after everyone else did. A landmark document, and basic text for any serious agricultural researcher.

Vandana Shiva. *The Violence of the Green Revolution*. Penang, Malaysia: Third World Network, 1991. How the work of Norman Borlaug and his colleagues went awry.

Vandana Shiva. *Monocultures of the Mind*. Penang, Malaysia: Third World Network, 1993. A description of the industrial agriculture mindset, as seen from the Third World.

Cynthia Barstow. *The Eco-Foods Guide: What's Good for the Earth Is Good for You.* Gabriola Island, B.C.: New Society Publishers, 2002. How to shop and eat healthily, with minimum contact with the horrors of industrial food.

EQUIPMENT AND TOOLS

Lehman's. *Heritage Non-Electric Catalog.* Kidron, Ohio: Lehman Hardware and Appliances, Inc., 2003. This is the mail-order catalog the Amish and the Mennonites use to outfit their traditional farms. One of the most fascinating catalogs in the world, filled with such items as woodstoves, hand fruit presses, and scythes, all at reasonable prices, it's one of the major sources for overseas development workers in Third World countries. It comes out every year.

To order a copy, write: Lehman's Hardware and Appliances, Inc., One Lehman Circle, P.O. Box 41, Kidron, Ohio 44636 USA. (888) 438-5346. Or go online at: (www.Lehmans.com).

Ken Darrow and Mike Saxenian. *Appropriate Technology Sourcebook.* Stanford, Appropriate Technology Project/Volunteers in Asia, 1993. An international successor to the *Whole Earth Catalog*, with some of the simplest, as well as most advanced high-tech instruments for work and for change.

Harris Pearson Smith. *Farm Machinery and Equipment.* New York: McGraw-Hill, 1965. The basic (now outdated) text I used when first learning about farm implements.

Food and Agriculture Organization of the United Nations. *Draught Animal Power Manual: A Training Manual for Use by Extension Agents.* Rome: Food and Agriculture Organization, 1994. Everything anyone ever wanted to know about using draft (or draught) animals on the small farm or homestead.

EXPOSÉS OF INDUSTRIAL AGRICULTURE

Karen Davis. *Prisoned Chickens, Poisoned Eggs.* Summertown, TN: Book Publishing Company, 1996. A graphic, and horrifying description of the modern, corporate poultry industry.

Gail A. Eisnitz. *Slaughterhouse: The Shocking Story of Greed, Neglect and Inhumane Treatment Inside the U.S. Meat Industry.* Amherst, N.Y.: Prometheus Books, 1997. The book's title says it all.

Alexander M. Ervin, Cathy Holtslander, Darrin qualman, Rick Sawa, eds. *Beyond Factory Farming: Corporate Hog Barns and the Threat to Public Health, the Environment and Rural Communities.* Saskatoon, Saskatchewan: Canadian Centre for Policy Alternatives, 2003. An exposé of the modern industrial hog farm and its horrors.

Frank Browning. *The Vanishing Land: The Corporate Theft of America.* New York: Harper Colophon Books, 1975. Dated, but accurate description of the corporate takeover of American farming.

Nicols Fox. *Spoiled: The Dangerous Truth About a Food Chain Gone Haywire.* New York: Basic Books, 1997. What you never wanted to know about the meat industry's dark underside.

Eric Schlosser. *Fast Food Nation.* New York: Perennial/Harper Collins Publishers, 2002. It may be fast, but it's not good. This bestseller tells why.

Wendell Berry. *The Unsettling of America: Culture and Agriculture.* San Francisco: Sierra Club Books, 1986. A farmer/poet's passionate case against corporate control of farming.

Harry P. Diaz, Joann Jaffe and Robert Stirling, eds. *Farm Communities at the Crossroads: Challenge and Resistance.* Regina, Saskatchewan:

Canadian Plains Research Centre, 2003. A comprehensive, multi-authored look at prairie rural communities.

MACROECONOMICS: THE GROWTH OF FACTORY FARMING AND THE WORLD FOOD INDUSTRY

Brewster Kneen. *From Land to Mouth: Understanding the Food System.* Toronto: NC Press Limited, 1995. The history, theory and crimes of modern agribusiness, as seen by Canada's foremost food industry critic.

A.V. Krebs. *The Corporate Reapers: The Book of Agribusiness.* Washington, D.C.: Essential Books, 1992. Krebs, publisher of the *Agribusiness Newsletter* and the *Calamity Howler*, is a dedicated defender of the family farm and a meticulous researcher. This is his analysis of the development and growth of U.S. agribusiness.

Walden Bello. *Dark Victory: The United States, Structural Adjustment and Global Poverty*. Penang, Malaysia: Third World Network, 1994. How the international food industry system keeps the poor poor, and the rich rich.

Tim Lang and Colin Hines. *The New Protectionism: Protecting the Future Against Free Trade.* London: Earthscan Publications, 1993. An economic analysis, by two unorthodox economists, of the new world trade system and agriculture's place in it.

Andrew Kimbrell, ed. *Fatal Harvest: The Tragedy of Industrial Agriculture.* Washington, D.C.: Island Press, 2002. An excellent, lavishly illustrated overview of the issues facing North Americans in the area of food and agriculture. The only problem with it is that Island Press decided to publish it as a massive, hardbound, coffee-table book, beyond the budget of most people, and far too heavy to carry from one place to another. Go figure.

MISCELLANEOUS, A.K.A. INTERESTING

Ralph Whitlock. *Rare Breeds.* New York: Van Nostrand Reinhold Company, 1980. An illustrated compendium of rare livestock breeds and those organizations dedicated to protecting them from extinction.

Janet Vorwald Dohner. *The Encyclopedia of Historic and Endangered Livestock and Poultry Breeds*. New Haven, CT: Yale University Press, 2001. An international, and more exhaustive, version of the previous title.

Brian Capon. *Botany for Gardeners: An Introduction and Guide*. Portland, Oregon: Timber press, 1990. For gardeners with a scientific bent, this book provides the botanical background for their amateur efforts.

Beryl Brintnall Simpson and Molly Conner Ogorzaly. *Economic Botany: Plants in Our World.* New York: McGraw Hill Higher Education, 2001 This will open the eyes of those who don't understand why nations have gone to war over plants.

Andrew Dalby, ed. *Cato On Farming: De Agricultura*. Blackawton, Totnes, Devon: Prospect Books, 1998. One of the earliest treatises on the agricultural arts, written for the ancient Romans by someone who loved the country. It is sometimes well for modern readers to realize that the ancients, even before the Caesars, had thought about and studied the art and science of growing food.

Henry David Thoreau. *Walden.* Koln: Konemann Versgesellschaft mbH, 1996. Thoreau's account of his year living alone on the shores of Walden Pond. It seems fitting, as well as ironic, that this edition of an American classic should have been printed in modern, democratic Germany, once home to the Nazis. In today's corporate America, Thoreau would likely have been arrested, stripped of his citizenship and shipped off to a cage in Guantanamo Bay, for defying the corporate "will to power."

Lee Allen Peterson. **Edible Wild Plants.** Boston: Houghton Mifflin, 1977. Updated and republished yearly, this is an accurate, authoritative field guide to edible plants in the U.S. and Canada, including many that were staples for First Nations people.

Ross & Linda Maracle. **First Nations Cooking: Creating a Family Heirloom.** Deseronto, Ontario: Epic Press, 2003. Recipes taken from the (previously) unwritten cookbooks of North America's native peoples.

Eleanor Noss Whitney and Sharon Rady Rolfes. **Understanding Nutrition.** Belmont, California: Wadsworth/Thomson Learning, 2002.

A university-level textbook, whose content is accessible to lay people. One of the best discussions of every aspect of human nutrition available anywhere in the English language. Unfortunately, like most hardbound textbooks, it is expensive.

ORGANIZATIONS AND CITIZENS' GROUPS

National Farmers Union (NFU): "is farm families sharing common goals ... [promoting] the family farm as the most appropriate and efficient means of agricultural production." There are branches in both the U.S. and Canada:

National Farmers Union, 400 North Capitol St. NW, Suite 790, Washington, D.C. 20001 USA (202) 554-1600. www.nfu.org

National Farmers Union, 2717 Wentz Ave., Saskatoon, Saskatchewan S7K 4B6 Canada (306) 652-9465. www.nfu.ca

National Family Farm Coalition "represents family farm and rural groups in 30 states whose members face the challenge of the deepening economic recession in rural communities caused primarily by low farm prices and the increasing corporate control of agriculture."

National Family Farm Coalition, 110 Maryland Ave. NE, Suite 307, Washington, D.C. 20002 USA (202) 543-5675, www.nffc.net.

Seed Savers Exchange: the oldest and largest U.S. heritage seed saving organization. Its annual catalog is a genuine treasure.

Seed Savers Exchange, 3076 North Winn Road, Decorah, Iowa 52101 USA, (563) 382-5990, www.seedsavers.org

Seeds of Diversity Canada: the Canadian equivalent of Seed Savers Exchange.

Seeds of Diversity Canada, P.O. Box 36, Station Q, Toronto, Ontario M4T 2L7 Canada, (905) 623-0353, www.seeds.ca.

Center for Science in the Public Interest: a consumer health advocacy organization with U.S. and Canadian branches, one of whose issues is truth in food labelling.

Center for Science in the Public Interest, 1875 Connecticut Ave. NW, Suite 300, Washington, D.C., 20009 USA, (202) 332-9110, www.cspinet.org

Centre for Science in the Public Interest, Suite 4550, CTTC Building, 1125 Colonel By Drive, Ottawa, Ontario K1S 5R1 Canada. (613) 244-7337, http://www.cspinet.org/canada.

City Farmer: Canada's office for urban agriculture. In business since the 1970s, this organization has a lot of expertise and great links to other sites.

City Farmer, Box 74561, Kitsilano, RPO, Vancouver, B.C. V6K 4P4 Canada. (604) 685-5832, www.cityfarmer.org.

Slow Food: you don't have to like Italian food to appreciate this group, dedicated as they are to fostering the good life.

Slow Food, Via della Mendicita Istruita 8, 12042 Bra (Cuneo), Italy. 39 0712-419611, www.slowfood.com.

Local Harvest: farmers markets and community supported agriculture farms in the USA.

www.localharvest.org

Native Seeds/SEARCH: a non-profit organization that works to conserve the traditional crops, seeds and farming methods of native peoples.

Native Seeds/SEARCH, 526 N 4th Ave., Tucson, Arizona 85705 USA, (520) 622-5561.

ETC Group: an international, civil society group headquartered in Canada and dedicated to the advancement of cultural and ecological diversity and human rights. Formerly known as the Rural Advancement Foundation (RAFI), it is particularly strong in its focus on agricultural issues.

ETC Group, 1 Nicholas St., Suite 200B, Ottawa, Ontario K1N 7B7 Canada, (613) 241-2267, www.etcgroup.org

CHAPTER 1

[1] Available for free download from the National Agricultural Library/USDA website, at www.nal.usda.gov. It is only 47 pages long, so it doesn't use up too much printer ink to make a hardcopy.

[2] Also available for download from the same NAL/USDA website mentioned in note 1. Be warned, however, that this is a massively long document, of which the vegetable section alone totals 1,520 pages.

[3] United States Department of Agriculture, Agricultural Research Service, *Agriculture Handbook # 8: Composition of Foods* (Washington, D.C.: USDA, 1963).

[4] Whitney, E. N. and Rolfes, S. R., *Understanding Nutrition* (Belmont, CA: Wadsworth Thomson Learning, 2002), 379.

[5] Watt, B. K. and Merrill, A. L., *Agriculture Handbook # 8, Composition of Food—Raw, Processed, Prepared* (Washington, D.C.: USDA, 1950).

[6] Whitney and Rolfes, *Understanding Nutrition*, 398.

[7] *Ibid.*

[8] *Ibid.*, 399.

[9] *Ibid.*

[10] For information, see their website at www.seedsavers.org.

[11] Whealy, Kent. Personal interview, May 2003.

[12] Hartz, Timothy. Personal interview, May 2003.

[13] Information taken from the California Tomato Growers Association web page at www.ctga.org.

[14] Information taken from the Florida Tomato Committee website at www.floridatomatoes.org.

[15] Florida Department of Agriculture, *Florida Agricultural Statistics, Vegetable Summary* (Orlando, FL; Florida Department of Agriculture, 2001). Downloaded from www.nass.usda.gov.

[16] Campil–Agro Industrial do Campo do Tejo, Ld., *Tomato Varieties* (Cartaxo, Portugal: Campil, 1999). Downloaded from www.campil.web.pt.

[17] University of California, Davis, Department of Vegetable Crops, California Tomato Research Institute, 2002 Annual Project Report, *2002 California Top 50 Varieties* (Davis, CA: California Tomato Research Institute, 2003), 2.

[18] Simonne, Eric, ANR 1143 "Tomato Varieties for Commercial Production." Alabama Cooperative Extension System (Auburn, AL: Alabama Department of Agriculture, 1998).

[19] Simpson, B. B. and Ogorzaly, M. C., *Economic Botany: Plants in Our World* (New York: McGraw Hill, 2001), 89.

[20] Maul, F., "Tomato Flavor and Aroma Quality as Affected by Storage Temperature." *Journal of Food Science* 65(7), Oct. 2000, 1228–1237.

CHAPTER 2

[1] Picard, Andre, "Today's Fruits, Vegetables Lack Yesterday's Nutrition," *The Globe and Mail*, July 6, 2002, A1.

[2] Mayer, Anne-Marie, "Historical Changes in the Mineral Content of Fruits and Vegetables," *British Food Journal* 99(6), 1997, 207–211.

[3] *Ibid.*, 209.

[4] Pettinato, John R., "Green-Eyed and Depressed," *Discover* 24(6), June 2003, 23–24.

[5] *Ibid.*, 24.

[6] *Ibid.*, 23.

[7] Whitney and Rolfes, *Understanding Nutrition*, 447.

[8] Holmes, H. Nancy, *Professional Guide to Diseases* (Springhouse, PA: Springhouse Corporation, 2001), 883.

[9] Whitney and Rolfes, *Understanding Nutrition*, 335.

[10] Picard, "Today's Fruits," A4.

[11] Burning is, chemically speaking, the combining of some substance or element with oxygen, during which heat and/or light energy is often given off as a by-product. When wood burns, it combines with oxygen, and produces heat, light, and ash. When iron rusts, it also combines with oxygen in a sort of "slow fire" that produces rust.

[12] Krogh, David, *Biology: A Guide to the Natural World* (Upper Saddle River, NJ: Prentice Hall, 2000), 26.

[13] *Ibid.*

[14] *Ibid.*

[15] Whitney and Rolfes, *Understanding Nutrition*, 358–9.

[16] *Ibid.*

[17] *Ibid.*

[18] *Ibid.*

[19] *Ibid.*, 434.

[20] *Ibid.*

[21] Kneen, Brewster, *From Land to Mouth: Understanding the Food System* (Toronto: NC Press Limited, 1995), 52.

[22] Kimbrell, Andrew, ed., *Fatal Harvest: The Tragedy of Industrial Agriculture* (Washington: Island Press, 2002), 79.

[23] *Ibid.*, 58.

[24] "The Harrowsmith Bread Test," *Harrowsmith* IV (4), December 1979, 34.

[25] Robertson, David, Moya Beal and Paul Schmidt, "Richest of the Enriched," *Harrowsmith* IV (4), December 1979, 32.

[26] Do, Sylvie, "38 marques sur le gril," *Protegez-vous*, August 2003, 9.

[27] *Ibid.* 8.

[28] *Ibid.* 11.

[29] *Ibid.*

[30] Picard, Andre, "Multivitamin Supplements a Fix for Food's Shortcomings, Experts Say," *The Globe and Mail*, July 6, 2002, A4.

[31] *Ibid.*

32 Whitney and Rolfes, *Understanding Nutrition*, 350.

33 Meikle, James, "Vitamin Pills Could Damage Health," *The Guardian*, May 8, 2003, online edition, www.guardian.co.uk/Print/0,3858,4663560,00.html.

34 *Ibid.*

35 "Vitamin E, Beta-Carotene Do Little for Heart: Study," The Associated Press, June 12, 2003, online edition, reprinted by *The Globe and Mail*, www.globeandmail.com/servlet/RTGAMArticleHTMLTe

36 *Ibid.*

37 *Ibid.*

38 Whitney and Rolfes, *Understanding Nutrition*, 350.

CHAPTER 3

1 Washington, Mary. "Pre-Seasoned Meat Could Be the Death of You." *The Westsider* 2(5), 2002: 2.

2 *Ibid.*

3 *Ibid.*

4 "Enhanced meat." *Virtual Weber Bullet* www.virtualweberbullet.com/enhanced-meat.html. June 23, 2003, 2.

5 *Ibid.*, 4.

6 Revill, Jo. "McDonald's Bows to Critics and Slashes Salt Ration." *The Observer* online edition www.observer.co.uk. March 6, 2004, 1–2.

7 Munro, Margaret. "Health Canada Keeps Riskiest Chip a Secret." *The National Post*, October 1, 2002, A2.

8 Kaufman, Marc. "FDA Finds Potential Cancer Agent in Fries." *The Washington Post*, December 5, 2002, A10.

9 Center for Science in the Public Interest (CSPI). "Chemical Cuisine: CSPI's Guide to Food Additives." www.cspinet.org/reports/chemcuisine.htm. January 19, 2001, 20.

10 Press Association. "Dangerous Dye Levels Found in Tikka," reprinted in *The Guardian*, online edition www.guardian.co.uk. March 23, 2004, 1–2.

11 Khoo, Michael, "Want Drugs with Those Fries?" *Tom Paine* online edition www.tompaine.com/feature.cfm/ID/7630/view/print. May 1, 2003, 1.

12 Burton, G.R.W. and Engelkirk, P. G. *Microbiology for the Health Sciences*, (Philadelphia: Lippincott Williams & Wilkins, 2000), 200.

[13] *Ibid.*

[14] Holmes, *Professional Guide to Diseases*, 165.

[15] *Ibid.*

[16] *Ibid.*, 195–6.

[17] Associated Press. "Antibiotic-Resistant Syphilis Spreading." *The Globe and Mail* online edition www.theglobeandmail.com/servlet/story/RTGAM.20040708. July 8, 2004, 1.

[18] Gibson, Richard. "Resistance to McDonald's New Antibiotic Feed Ban Policy." Dow Jones Newswires (as quoted in *The Agribusiness Examiner*, No. 261, June 24, 2003, 7).

[19] Canadian Press. "Superbug: Nightmare Now Reality," as reprinted in the *Globe and Mail*, July 8, 2002, A1.

[20] *Ibid.*

[21] Kirkey, Sharon, "Antibiotics Not Good for Infants" *Regina Leader-Post*, October 1, 2003, A1.

[22] Meikle, James, "Danger Warning after Increase in Drug Residues Found in Eggs," *The Guardian* online edition www.guardian.co.uk. April 30, 2004, 1.

[23] Netscape Home & Real Estate page channels.netscape.com/ns/home-realestate/package.jsp?name=f. January 28, 2004

[24] *Ibid.*

[25] Hansen, Michael, Jean M. Halloran, Edward Groth III, Lisa Leferts. "Potential Public Health Impacts of the Use of Recombinant Bovine Somatotropin in Dairy Production." Joint WHO/FAO Expert Committee on Food Additives, September 1997, as posted on the U.S. Consumers Union website www.consumersunion.org/food/bgh-codex.htm., 7.

[26] *Ibid.*

[27] *Ibid.*, 9.

[28] *Ibid.*, 2.

[29] *Ibid.*, 5

[30] *Ibid.*, 12.

[31] *Ibid.*, 12-13.

[32] *Ibid.*

33 Wright, George. "Coca-Cola Withdraws Bottled Water from the U.K.'" *The Guardian* online edition www.guardian.co.uk. March 19, 2004, 1–2.

34 Mittelstaedt, Martin. "Contaminants in Gull Eggs Raising alarm," Toronto *Globe and Mail* online edition www.theglobeandmail.com/servlet/story/RTGAM.20040507 May 7, 2004, 1–2.

35 Mittelstaedt, Martin. "Contaminants in Gull Eggs Raising Alarm," Toronto *Globe and Mail* online edition www.theglobeandmail.com/servlet/story/RTGAM.20040507. May 7, 2004, 1–2.

36 Venes, Donald, editor, *Taber's Cyclopedic Medical Dictionary* (Philadelphia: F. A. Davis Company, 2001), 601.

37 Canadian Press. "Dioxins Taint Food Samples," *Regina Leader-Post*, September 16, 2002, A1.

38 *Ibid.*

39 *Ibid.*

40 Ontario Ministry of the Environment, *Guide to Eating Ontario Sport Fish 2001–2002,*" 21st edition, (Toronto: Queen's Printer for Ontario, 2001).

41 Picard, Andre. "Salmon: A Slippery Subject," Toronto *Globe and Mail* online edition www.theglobeandmail.com/servlet/story/RTGAM.20040116. January 16, 2004, 1–5.

42 Picard, Andre. "Girls Warned to Cut Back on Meat, Whole Milk," *The Globe and Mail*, online edition, www.sympatico.globeandmai...M&site=Front&configLabel=front&hub=Front. July 13, 2003, 1.

43 *Ibid.*

44 *Ibid.*

45 Turning Point Project, "Unlabelled, Untested and You're Eating It," advertisement No. 2 in a series on genetic engineering, October 14, 1999.

46 *Ibid.*

47 Ho, Mae-Wan. "GM DNA in Human Gut Underestimated," Institute of Science in Society (ISIS) Report July 21, 2002, 1–5.

48 Iyer, Vik. "Crop Gene 'Could Weaken Medicines,'" *Press Association (PA) News*, August 16, 2002. 1–2.

49 Rebecca Goldburg. "Pause at the Amber Light," *Ceres* 27(3), 1995, 21.

50 Christpeels, Maarten J., and Sadava, David E., *Plants, Genes, and Agriculture* (London: Jones and Bartlett Publishers, 1994) 423.

51 Weiss, Rick. "Insect Bambi Threatened by Gene-Altered Corn," *The Ottawa Citizen*, May 20, 1999, A13.

52 Whitney and Rolfes, *Understanding Nutrition*, 649.

53 *Ibid.*

54 Holmes, *Professional Guide to Diseases*, 1304.

55 Venes, *Taber's Cyclopedic Medical Dictionary*, 502.

56 *Ibid.*

57 Parker-Pope, Tara. "Canadian Mad Cow Discovery Exposes U.S. Beef Industry's 'Dirty Little Secret,'" *The Wall Street Journal*, as posted in *Agribusiness Examiner #252*, February 6, 2003.

58 CTV News. Bell Globemedia Inc., online edition www.ctv.ca. June 5, 2003, 1.

59 Johnson, Carleen. "Mad Cow Scare Now Hurting State Potato Industry," KOMO TV online edition www.komotv.com/news/printstory.asp?id=29056. January 8, 2004, 1.

60 Meikle, James. "Vets Investigate Mystery Brain Disease in Cattle," *The Guardian* online edition www.www.guardian.co.uk. June 8, 2004, 1–2.

61 Mittelstaedt, Martin. "Canada's Food Rich in Heavy Metals, Group Says," *The Globe and Mail*, May 5, 2003, A1.

62 *Ibid.*

63 Whitney and Rolfes, *Understanding Nutrition*, 451.

64 Harding, Anne. "Hold the Tuna," *Grist* magazine, reprinted in *Tom Paine*, online edition, April 2, 2004, 1.

65 Lawrence, Felicity. "Beef and Pork Proteins Found in Imported Chicken," *The Guardian*, online edition, www.guardian.co.uk/Print/0,3858,4621718,00.html. March 10, 2003, 1.

66 *Ibid.*

67 *Ibid.*

68 Rosenfeld, Steven. "Body Burden," *Tom Paine*, online edition, www.tompaine.com/feature.cfm/ID/7353/view/print. June 3, 2003 1.

69 *Ibid.*

70 *Ibid.*

71 *Ibid.*, 3.

[72] *Ibid.*, 1.

[73] Stevenson, Mark, "Tests Find Drug Taint in Water," *The Globe and Mail*, October 21, 2002, A1.

[74] Webster's Dictionary Online www.webster-dictionary.org/definition/nanotechnology. July 9, 2004.

[75] Merkle, Ralph. "Nanotechnology," Zyvex website www.zyvex.com/nano. July 9, 2004, 1.

[76] ETC Group discussion list. "Tenth Toxic Warning: More Evidence to Support Nano-Moratorium," April 1, 2004.

[77] ETC Group discussion list. "Jazzing Up Jasmine: Atomically Modified Rice in Asia?" March 25, 2004.

[78] Commission de police du Quebec (CECO), *Rapport Officiel: la Lutte au Crime Organise* (Montreal: Stanke, Editeur officiel du Quebec, 1976) 301–2.

[79] Canadian Press, "Beef Recall Ordered in Ontario," reprinted in *The Globe and Mail* online edition, August 25, 2003, 1–2.

[80] Canadian Press, "Ontario Meat Plant Shut Down," reprinted in *The Globe and Mail* online edition, October 8, 2003, 1.

[81] Robert Benzie, "Food Safety Overhaul Urged," *Toronto Star* online edition, July 23, 2004, 1.

[82] Canadian Press, "E. Coli Fear Prompts Massive Beef Recall," reprinted in *The Globe and Mail* online edition, August 8, 2004, 1.

[83] Jane Armstrong, "Alert Issued about Meat from Pickton's Pig Farm," *The Globe and Mail*, March 11, 2004, 1A.

[84] Fox, Nicols, *Spoiled: The Dangerous Truth About a Food Chain Gone Haywire* (New York: Basic Books, 1997).

[85] Eric Schlosser, *Fast Food Nation* (New York: Perennial, 2002) 197.

[86] Associated Press, "Fresh Fruits and Veggies New Frontier of Poisoning," reprinted in *The Globe and Mail* online edition, August 2, 2004, 1.

[87] *Ibid.*

[88] Witten, Mark, "On the Healthy Promise of Organic Foods," *CSL*, May/June 2003, 37.

[89] *Ibid.*

[90] Harvard University Department of Nutrition, "Trans Fatty and Coronary Heart Disease," www.hsph.harvard.edu/reviews/transfats.html. July 13, 2004, 1–16.

[91] The *Globe and Mail*. "What to do about Trans Fatty Acids?" *The Globe and Mail* online edition, www.theglobeandmail.com/servlet/ArticleNews/TPPrint/LA. December 11, 2003, 1–4.

[92] Harvard University Department of Nutrition, "Trans Fatty Acids and Coronary Heart Disease," 1–16.

CHAPTER 4

[1] FAO, *The State of Food and Agriculture* (SOFA) 1994 (Rome: FAO, 1994).

[2] Berry, Wendell, *The Unsettling of America* (San Francisco: Sierra Club Books, 1977).

[3] Collinson, Mike, "Green Evolution," *Ceres* 27(4), 23–24.

[4] Conway, Gordon R. and Pretty, Jules N., *Unwelcome Harvest: Agriculture and Pollution* (London: Earthscan Publications, 1991), 1.

[5] Mozafar, A., *Plant Vitamins: Agronomic, Physiological, and Nutritional Aspects* (Boca Raton, FL: CRC Press, 1994), iv.

[6] Lockeretz, William, ed. *Agricultural Production, and Nutrition* (Medford, MA: School of Nutrition Science and Policy, Tufts University 1997).

[7] Tremblay, N., ed. *Toward Ecologically Sound Fertilizer Strategies for Field Vegetable Production*, proceedings of the XXVI International Horticultural Congress (reprinted in *Acta Horticulturae* 627, October 2003).

[8] Rubin, Sandra, "Feeding Class Actions: Are Obesity Suits the Fodder?" *Financial Post* September 24, 2003, 12.

[9] For a thorough discussion of this problem, see Ernest Sternglass, *Secret Fallout: Low-Level Radiation from Hiroshima to Three-Mile Island* (New York: McGraw Hill, 1981) and Harvey Wasserman and Norman Solomon, *Killing Our Own: the Disaster of America's Experience with Atomic Radiation* (New York: Delta/Dell Publishing Co. 1982).

[10] Soil Improvement Committee, California Fertilizer Association, *Western Fertilizer Handbook: Second Horticulture Edition* (Danville, IL: Interstate Publishers Inc., 1998) 106.

[11] *Ibid*, 107.

[12] Mozafar, *Plant Vitamins*., 168-9.

[13] *Ibid*.

[14] *Ibid*., 169-171.

[15] *Ibid*., 171.

[16] *Ibid*., 172.

[17] *Ibid*, 173.

[18] *Ibid*, 174.

[19] *Ibid*, 186.

[20] *Ibid*, 187.

[21] *Ibid*, 196.

[22] Joji Muramoto, "Comparison of Nitrate Content in Leafy Vegetables from Organic and Conventional Farms in California," *Research Reviews*, Center for Agroecology and Sustainable Food Systems, University of California at Santa Cruz, Summer 2000 (8), 23-25.

[23] Mozafar, *Plant Vitamins*, 199

[24] Raven, Peter H., Evert Ray F., and Eichhorn, Susan E., *Biology of Plants*, 6th edition (New York: W. H. Freeman and Company, Worth Publishers, 1999), 727.

[25] *Ibid*.

[26] Mozafar, *Plant Vitamins*, 175.

[27] *Ibid*.

[28] *Ibid*.

[29] *Ibid*.

[30] Lockeretz, *Agricultural Production.*, 98.

[31] *Ibid*, 100.

[32] *Ibid*, 155.

[33] Tremblay, *Toward Ecologically*, 28.

[34] *Ibid*.

[35] Similarly exploitative attitudes to the braceros were common in other agricultural states, such as Michigan, whose annual cherry harvest was accomplished largely by Mexican workers. I still recall, as a young student in pre-veterinary medicine at Michigan State University, listening to the boastful tales of an agriculture student from Traverse City (the "cherry capital") of how he and his buddies used to gang-rape young Mexican girls in his father's barn. "If they told anybody, we'd just call the INS and have them kicked back to Mexico," he said laughing..

[36] Mississippi State University, College of Agricultural and Life Science, Agro-Ecosystem Information Systems www.ais.misstate.edu/AEE/2613/cases/diffusion-case.html. Case Study #2, 1-5.

[37] Brewer, Harold, "Give us this day," VDARE.com (Center for American Unity), www.vdare.com/misc/archive00/mechanization.htm. November 1, 2000, 1-3.

[38] Mississippi State University, Case Study # 2.

[39] Mines, Richard, "What Kind of Transition is Necessary to Secure the Future of U.S. Fruit, Vegetable and Horticultural Agriculture?" *Labor Management Decisions*, 8 (1), Winter–Spring 1999, 3.

[40] Gould, Wilbur A., *Tomato Production, Processing & Technology* (Timonium, MD: CTI Publications, 1992), 104–5.

[41] Mines, "What Kind of Transition," 3.

[42] Gould, *Tomato Production*, 104.

[43] Mozafar, *Plant Vitamins*, 253.

[44] Marathon Products Inc., "About Ethylene Gas," www.marathonproducts.com/products/ethyover.html. May 14, 2003, 1.

[45] Kader, Adel E. ed., *Postharvest Technology of Horticultural Crops* (Oakland, California: University of California Agriculture and Natural Resources Communication Services, Publication 3311, Third Edition, 2002), 150–1.

[46] Extoxnet Extension Technology Network, "Ethephon," pmep.cce.cornell.edu/profiles/extoxnet/dienochlor-glyphosate. August 28, 2004, 1.

[47] *Ibid* , 4.

[48] Mozafar, *Plant Vitamins*, 253.

[49] *Ibid*, 255.

[50] David Carle, *Introduction to Water in California* (Berkeley: University of California Press, 2004), 147.

[51] *Ibid*, 3–4.

[52] Mozafar, *Plant Vitamins*, 291–2.

[53] *Ibid*, 296–7.

[54] Carle, *Introduction to Water*, 149–50.

[55] Mozafar, *Plant Vitamins*, 298–9.

[56] *Ibid*, 300.

[57] Hartz, T. K., "Sustainable Vegetable Production in California: Current Status and Future Prospects," *Horticultural Science* 37(7), December 2002, 1017.

[58] Barstow, Cynthia, *The Eco-Foods Guide* (Gabriola Island, B.C.: New Society Publishers, 2002), 13.

[59] *Ibid*, 15.

[60] Hartz, "Sustainable Vegetable Production," 1017

[61] Ontario Ministry of Natural Resources, *Guide to Eating Ontario Sport Fish* (Toronto: Queen's Printer for Ontario, 2001).

[62] Barstow, *The Eco-Foods Guide,* 14.

[63] *Ibid.*, 15.

[64] Davis, Karen, "The Battery Hen: Her Life Is Not for the Birds," all-creatures.org homepage www.all-creatures.org/articles/egg-battery.html. September 21, 2004, 1–4.

[65] Davis, Karen, *Prisoned Chickens Poisoned Eggs: An Inside Look at the Modern Poultry Industry* (Summertown, TN: Book Publishing Company, 1996).

[66] Centers for Disease Control and Prevention (CDC), "Transmission of Influenza A Viruses Between Animals and People," as posted on CDC website, www.cdc.gov/flu/avian/gen- info/spread.htm. September 28, 2004.

[67] BBC News, "Avian Flu 'Discovered in Pigs,'" BBC News website, www.newsvote.bbc.co.uk/mpapps/pagetools/print/news.bbc.co.uk/1/. September 28, 2004, 1.

[68] Rick Dove, "The American Meat Factory," in *Beyond Factory Farming: Corporate Hog Barns and the Threat to Public Health, the Environment, and Rural Communities*, ed. Alexander M. Ervin et al. (Saskatoon: Canadian Centre for Policy Alternatives, 2003) 65–6.

[69] Eisnitz, Gail A., *Slaughterhouse: The Shocking Story of Greed, Neglect, and Inhumane Treatment Inside the U.S. Meat Industry* (Amherst, NY, Prometheus Books, 1997), 41-3.

[70] *Ibid.*

[71] *Ibid.*, 45-6.

[72] Jo Robinson, "Health Benefits of Grass-Fed Products," Eat Wild website, www.eatwild.com September 21, 2004.

[73] *Ibid.*

[74] Whitney and Rolfes, *Understanding Nutrition,* 144–6.

[75] D. S. Siscovick, T. E. Raghunathan, et al. "Dietary Intake and Cell Membrane Levels of Long-Chain n–3 Polyunsaturated Fatty Acids and the Risk of Primary Cardiac Arrest," *Journal of the American Medical Association* 274(17), 1363–7.

[76] Whitney and Rolfes, *Understanding Nutrition,* 586.

[77] Robinson, "Health Benefits."

[78] Duckett, S. K., Wagner, D. G., et al. "Effects of Time on Feed on Beef Nutrient Composition," *Journal of Animal Science* 71(8), 2079–88.

[79] Aro, A., Mannisto, S., Salminen, I., Ovaskainen, M.L., Kataja V., and Uusitupa, M. "Inverse Association Between Dietary and Serum Conjugated Linoleic Acid and Risk of Breast Cancer in Postmenopausal Women," Nutrition and Cancer 38, No. 2(2000), 151–7.

[80] Long, Cheryl and Keiley, Lynn "Is Agribusiness Making Food Less Nutritious?" *Mother Earth News* website www.motherearthnews.com/additional/print.php?id=2132 September 21, 2004, 1

[81] Gitte Meyer, "Pharma-Foods are Queuing up for Approval," *Eurosafe Newsletter*, 1–2/2003, as posted on the Centre for Bioethics and Risk Assessment website, www.bioethics.kvl.dk.

CHAPTER 5

[1] Not his real name. This anecdote was previously reported in Thomas Pawlick, *The Invisible Farm* (Chicago: Burnham Inc., 2001), 41–3. Burnham declared bankruptcy before the book could be distributed, and it is now out of print.

[2] Under Canada's "supply-management" farm marketing system, commodity marketing boards, such as the Ontario Milk Marketing Board or the Canadian Egg Marketing Agency, grant farmer/members "licenses"—referred to as quota—to sell specific amounts of a given board-regulated farm product. Quota can be bought and sold, and frequently changes hands when a farmer dies or retires.

[3] Conway and Pretty, *Unwelcome Harvest*, 2.

[4] *Ibid* 7.

[5] Epidemiological statistics from Vietnam, and detailed descriptions of the physiochemical processes by which dioxin affects living organisms, can be examined in Westing, Arthur H., *Herbicides in War: The Long-Term Ecological and Human Consequences* (New York, Peace Studies, 1984).

[6] Conway and Petty, *Unwelcome Harvest.*, 45.

[7] *Ibid.*, 37.

[8] James Hughes (ed.). *The Larousse Desk Reference* (New York: Larousse Kingfisher Chambers Inc., 1995), 81.

[9] Conway and Petty, *Unwelome Harvest*, 27.

[10] Repetto, R. *Paying the Price: Pesticide Subsidies in Developing Countries* (Washington: World Resources Institute, 1985).

[11] Switzer-Howse, K. D. and Coote, D. R."Agricultural Practices and Environmental Conservation," Ottawa: Agriculture Canada, 1984, 12.

12 Conway and Pretty, *Unwelcome Harvest*, 198.

13 Switzer-Howse and Coote, "Agricultural Practices," 12.

14 Taiganides, Paul E., "The Animal Waste Disposal Problem," *Agriculture and the Quality of Our Environment*, ed. Nyle C. Brady (Washington, D.C.: American Association for the Advancement of Science, 1967), 389–90.

15 *Ibid.*

16 Switzer-Howse and Coote, "Agricultural Practices," 13.

17 Linda Kane, "Swine Farm Neighbors Say Stink Bugs Them," *Amarillo Globe* website www.amarillo.com/stories/052898/new_020-3719.shtml. May 28, 1998.

18 Bill Paton, "The Smell of Domestic Pig Production on the Canadian Prairies," in *Beyond Factory Farming: Corporate Hog Barns and the Threat to Public Health, the Environment, and Rural Communities*, ed. Alexander M. Ervin et al. (Saskatoon: Canadian Centre for Policy Alternatives, 2003) 80–1.

19 A. Dennis McBride, State Health director, "Medical Evaluation & Risk Assessment: The Association of Health Effects with Exposure to Odors from Hog Farm Operations," North Carolina Department of Health and Human Services, 7 December 1998, as posted at website www.epi.state.nc.us/epi/mera/ilodoreffects.html.

20 *Ibid.*, 4.

21 Rick Dove, "The American Meat Factory," 60.

22 *Ibid.*

23 Natalie James, "Report Arouses Concern Over Government Role in *E. Coli* Deaths," *The Canadian Press*, July 28, 2000.

24 Dove, "The American Meat Factory," 61.

25 For an in-depth discussion of these patterns and their effects on the environment, see Lennart Hansson, Lenore Fahrig, and Gray Merriam (Eds.), *Mosaic Landscapes and Ecological Processes* (New York: Chapman and Hall, 1995).

26 O'Connor, Raymond J. and Shrubb, Michael. *Farming and Birds* (Cambridge: Cambridge University Press, 1986) 80.

27 *Ibid.*, 149.

28 *Ibid.*, 83.

29 *Ibid.*, 83-84.

30 *Ibid.*

31 *Ibid.*, 86.

[32] Elgie, Stewart, "Bite: Endangered Species Need a Law That Gives Them Some," *Environment Views* 18(1), 21.

[33] Wylynko, David. "The Rate Debate: Will the End of the Transportation Subsidy for Prairie Wheat Lead to More Sustainable Farming Practices on the Great Plains?" *Nature Canada* 25(1), 17–21.

[34] Van Tighem, Kevin, "Save the Gopher," *Environment Views* 18(1), 16–19.

[35] Edward B. Barbier et al., *Elephants, Economics and Ivory* (London: Earthscan Publications Ltd., 1990), 14.

[36] *Ibid.,* 17.

[37] Pitman, Dick, "Wildlife as a crop," *Ceres* 22(1), 30.

[38] *Ibid.,* 30–35.

[39] *Ibid.,* 35.

[40] R.B. Martin, "A Voice in the Wilderness," *Ceres* 26(6), 26, 27.

[41] *Ibid.,* 26.

[42] *Ibid.*

[43] Switzer-Howse and Coote, "Agricultural Practices," 22.

[44] *Ibid.*

[45] *Ibid.,* 8.

[46] *Ibid.,* 23.

[47] *Ibid.,* 24.

[48] Carle, *Introduction to Water,* 149.

[49] *Ibid.,* 150.

[50] *Ibid.,* 151.

[51] *Ibid.,* 152-3.

[52] *Ibid.,* 153.

[53] *Ibid.,* 155.

[54] Postel, Sandra, "Waters of Strife," *Ceres* 27(6), 19.

[55] *Ibid.,* 20.

[56] *Ibid.,* 21.

[57] *Larousse Desk Reference,* 146.

58 Postel, "Waters of Strife," 23.

59 Appelgren, B. and Burchi, S., "The Danube's Blues," *Ceres* 27(6), 24–28.

60 Braun, Armelle, "The Megaproject of Mesopotamia," *Ceres* 26(2), 25–30.

61 Carle, *Introduction to Water,* 115.

62 Stroud, Polly, "Africa's Wave of the Future, or a Backwash from the Past?" *Ceres* 27(4), 37.

63 Baeza-Lopez, Patricia, "A New Plant Disease: Uniformity," *Ceres,* 26(6), 41.

64 Pawlick, Thomas, "The Cause and Its Effects," *Harrowsmith* VII (2), 35.

65 *Ibid.*

66 Baeza-Lopez, "A New Plant Disease," 44.

67 Shiva, Vandana, "Mistaken Miracles," *Ceres,* 27(4), 28–29.

68 *Ibid.,* 29–30.

69 Stroud, "Africa's Wave of the Future," 39.

70 Herbert, John, "The Narrowing of the Options," *Ceres,* 26(6), 42–45.

71 Goldburg, Rebecca, "Pause at the Amber Light," *Ceres,* 27(3), 21.

72 Alfred W. Crosby, *Ecological Imperialism: The Biological Expansion of Europe 900–1900* (Cambridge: Cambridge University Press, 1986), 75.

73 Larousse, *Desk Reference,* 126, 136.

74 Crosby, *Ecological Imperialism,* 154–55.

75 Goldburg, "Pause at the Amber Light," 23.

CHAPTER 6

1 For an in-depth discussion of Russian agriculture up to and including the early communist era, see Kingston-Mann, Esther and Mixter, Timothy eds., *Peasant Economy, Culture and Politics of European Russia, 1800–1921* (Princeton: Princeton University Press, 1991).

2 Pawlick, "The Cause and Its Effects," 29.

3 Krebs, A. V., *The Corporate Reapers: the Book of Agribusiness* (Washington, D.C.: Essential Books, 1992), 237.

4 Kneen, Brewster, *From Land to Mouth: Understanding the Food System* (Toronto: NC Press, 1995), 70.

5 *Ibid.*

[6] *Ibid.*

[7] A good description of the situation in Third World cities is provided in Hardoy, Jorge, Mitlin, Diana, and Satterthwaite David, *Environmental Problems in Third World Cities* (London, Earthscan Publications, 1993).

[8] *Ibid.*

[9] Kotz, Nick, "Agribusiness," in *Radical Agriculture,* ed. Richard Merrill (New York: New York University Press, 1976), 41–51.

[10] *Ibid.*, 48.

[11] *Ibid.*, 49.

[12] Turk, Linda, "No Sacred Cows in the NAFTA Era," *The Globe and Mail,* April 25, 1996: A20.

[13] *Ibid.*

[14] Fagan, "Canada Defends Farm Taridds," *The Globe and Mail,* April 30, 1996: A1–2.

[15] Pawlick, *The Cause and Its Effects,* 4.

[16] Berry, Wendell, *The Unsettling of America* (San Francisco: Sierra Club Books, 1977), 40.

[17] *Ibid.*, 41.

[18] *Ibid.*, 43.

[19] *Ibid.*, 47.

[20] *Ibid.*, 9.

[21] *Ibid.*, 6.

[22] Pawlick, *The Cause and Its Effects,* 29.

[23] *Ibid.*, 28.

[24] Terry Pugh, "Thousands of Family Farmers Will Become Casualties to this Adjustment," *Ceres,* 27(1), 1995, 28–33.

[25] Krebs, *The Corporate Reapers,* 230.

[26] *Ibid.*

[27] *Ibid.*, 231.

[28] *Ibid.*, 247.

[29] Tangermann, Stefan, "A Major Step in a Good Direction," *Ceres* 27(1), 24–27.

[30] *Ibid.*

31 Lang, Tim and Hines, Colin, *The New Protectionism*, (London: Earthscan Publications, 1993).

32 Lang, Tim and Hines, Colin, "A Disaster for the Environment, Rural Economies, Food Quality, and Food Security," *Ceres*, 27(1), 19–23.

33 Canadian Press, "U.S. Gears Up for Food Fight with Canada," *The Ottawa Citizen* January 30, 1996: A7, and Fagan, "Canada Defends Farm Tariffs," A1–2.

CHAPTER 7

1 The Rodale Institute, publishers of *Organic Gardening* magazine, was founded by Robert Rodale and has continued under his son, J. I. Rodale. It is one of the world's leading research facilities for organic horticulture and sustainable agriculture.

2 Stillman, Janice, ed., *The Old Farmers' Almanac* (Dublin, NH: Yankee Publishing Inc.).

3 Geiger, Peter and Duncan, Sondra, eds., *Canadian Farmers' Almanac* (Lewiston, ME: Almanac Publishing Company).

4 Rural Advancement Foundation International www.rafi.org, "Earmarked for Extinction? Seminis Eliminates 2,000 Varieties," July 21, 2000.

5 *Ibid.*

6 Contact details:
• Seed Savers Exchange, Inc., 3076 North Winn Road, Decorah, Iowa 52101 USA; (563) 382-5990; www.seedsavers.org.
• Seeds of Diversity Canada, P.O. Box 36, Station Q, Toronto, Ontario M4T 2L7 Canada; (905) 623-0353; www.seeds.ca

7 Ashworth, Suzanne, *Seed to Seed: Seed Saving and Growing Techniques for Vegetable Gardeners* (Decorah, IA: Seed Savers Exchange, 2002).

8 *Southern Exposure Seed Exchange Catalogue and Garden Guide* (Mineral, Virginia: Southern Exposure Seed Exchange, 2002).

9 *Seeds of Change Annual Seed Book 2001* (Santa Fe, NM: Seeds of Change, 2001).

10 Portland Community Gardens, 6437 SE Division, Portland, OR 97206, USA; (503) 823-1612; www.parks.ci.portland.or.us/Parks/ CommunityGardens.htm.

11 City Farmer—Canada's Office of Urban Agriculture, Box 74561, Kitsilano RPO, Vancouver, B.C. V6K 4P4, Canada; (604) 685-5832; www.cityfarmer.org.

12 Farmers Markets of Ontario, www.farmersmarketontario.com.

13 Local Harvest, www.localharvest.org.

14 Katie Zezima, "Challenging Mini-Marts and Large Supermarket Chains, Vermont

Country Stores Organize," *New York Times,* as posted in *Agribusiness Examiner,* www.ea1.com/CARP Issue #382, November 30, 2004, 1.

[15] Associated Press, "Nearly Extinct Turkey Breeds Brought Back to Life, Table," posted at the Toronto *Globe and Mail* online edition, www.theglobeandmail.com/servlet/story/RTGAM.20041124. November 24, 2004, 1.

[16] Dohner, Janet Vorwald, *The Encyclopedia of Historic and Endangered Livestock and Poultry Breeds* (New Haven: Yale University Press, 2001), 3–5.

[17] Helen and Scott Nearing abandoned city life to homestead in rural Vermont in 1932. *Living the Good Life: How to Live Sanely and Simply in a Troubled World* was their first book, written after they'd spent nearly 20 years on the farm, and became a "cult" bestseller. Several other titles followed, chronicling the couple's further rural adventures. *The Good Life* (ISBN 0805209700), a reprint edition combining *Living the Good Life* and *Continuing the Good Life,* was published in New York by Schocken Books in 1990.

CHAPTER EIGHT

[1] Thom Hartmann, "Corporations Are People Too," *Tom Paine Magazine* online edition, www.tompaine.com/feature.cfm/ID/7603/view/. April 18, 2003, 1–6.

[2] One of the earliest, and still unsurpassed, exposes on this subject was Green, Mark J., Fallows, James M., and Zwick, David R., *Who Runs Congress? The President, Big Business or You?* (New York: Bantam/Grossman, 1972.)

[3] See especially: Krebs, A.V., *The Corporate Reapers: The Book of Agribusiness* (Washington, D.C.: Essential Books, 1992), and Kneen, Brewster, *From Land to Mouth: Understanding the Food System* (Toronto: NC Press, 1995).

[4] Russel, Sabin, "FDA Lax in Drug Safety, Journal Warns," *San Francisco Chronicle* online edition www.sfgate.com/cgi-bin/article.cgi?file=/c/a/2004/11/23/MN, 1.

[5] ETC Group, "Maize Rage in Mexico: GM Maize Contamination in Mexico—Two Years Later," article posted to the ETC Group discussion list, October 10 2003, 1.

[6] "GM Trials Reveal Mixed Impact on Wildlife," *The Guardian* online edition, www.guardian.co.uk/print/0,3858,4776002-103528,00.html, October 16, 2003.

[7] Novotny, Eva, "Defending Nature is Not Anti-Science," *The Guardian* online edition, www.guardian.co.uk/print/0,3858,4791089-103677,00.html, November 6, 2003, 2.

[8] Weise, Elizabeth, "Modified Crops Issue Divides Voters," *USA Today* online edition, usatoday.printthis.clickability.com/pt/cpt?action=cpt&title=U, November 4, 2004, 1.

[9] *Ibid.*

[10] Elizabeth Rosenthal, "Europe Closes Ranks on Bioengineered Food," *International Herald Tribune* online edition, www.iht.com/bin/print.php?file=542067.html, October 5, 2004, 1.

[11] Pratley, Nils. "Kick Them Where it Hurts," *The Guardian* online edition, www.guardian.co.uk, December 18, 2003, 1.

[12] Kimbrell, Andrew, ed., *Fatal Harvest: The Tragedy of Industrial Agriculture* (Washington, D.C.: Island Press, 2002), 59.

CHAPTER NINE

[1] Waters, Alice, *Chez Panisse* website, www.chezpanisse.com/alice.html, September 10, 2004.

[2] Margulis, Lynn, *Origin of Eukaryotic Cells.* (New Haven: Yale University Press, 1970.) and *Symbiotic Planet: A New Look at Evolution.* (New York: Basic Books, 1998.)

[3] Pollan, Michael, "Cruising on the Ark of Taste," *Mother Jones* online edition, www.mojones.com/cgi-bin/print_art.com/news/feature/2003/19/ma_372_01.html, June 3, 2003, 1.

[4] Gibbons, Euell, *Stalking the Wild Asparagus* (New York: David McKay Co., 1962).